STRUCTURAL FOAM

PLASTICS ENGINEERING

Series Editor

Donald E. Hudgin
Princeton Polymer Laboratories
Plainsboro, New Jersey

1. Plastics Waste: Recovery of Economic Value, *Jacob Leidner*
2. Polyester Molding Compounds, *Robert Burns*
3. Carbon Black-Polymer Composites: The Physics of Electrically Conducting Composites, *Edited by Enid Keil Sichel*
4. The Strength and Stiffness of Polymers, *Edited by Anagnostis E. Zachariades and Roger S. Porter*
5. Selecting Thermoplastics for Engineering Applications, *Charles P. MacDermott*
6. Engineering with Rigid PVC: Processability and Applications, *Edited by I. Luis Gomez*
7. Computer-Aided Design of Polymers and Composites, *D. H. Kaelble*
8. Engineering Thermoplastics: Properties and Applications, *edited by James M. Margolis*
9. Structural Foam: A Purchasing and Design Guide, *Bruce C. Wendle*
10. Plastics in Architecture: A Guide to Acrylic and Polycarbonate, *Ralph Montella*

Other Volumes in Preparation

STRUCTURAL FOAM

A Purchasing and Design Guide

BRUCE C. WENDLE
Celpro, Inc.
Kent, Washington

MARCEL DEKKER, INC. New York and Basel

Library of Congress Cataloging in Publication Data

Wendle, Bruce C.
Structural foam.

(Plastics engineering ; 9)
Includes index.
1. Plastic foams. I. Title. II. Series: Plastics engineering (Marcel Dekker, Inc.) ; 9.
TP1183.F6W46 1985 668.4'93 85-1548
ISBN 0-8247-7398-5

MARCEL DEKKER, INC.
270 Madison Avenue, New York, New York 10016

Current printing (last digit):
10 9 8 7 6 5 4 3 2 1

PRINTED IN THE UNITED STATES OF AMERICA

Preface

Over the years since plastic polymers were first introduced, the industry has continually developed new ways to process, form, and utilize various types of materials. Injection molding, extrusion, vacuum forming, and blow molding are but a few of the ways one can produce a plastic product. Most of these techniques involve solid polymers. Expanded polystyrene and foamed board stock were later variations.

In the late 60s Union Carbide, working on a modified molding process, coined the expression "structural foam." While not a very definitive name it stuck, and for better or worse we have built a relatively small but active industry around it. Is structural foam a material form or a process? Described sometimes as "just another club in the bag," it offers many advantages and must be considered as a viable means of producing a plastic product.

Several books on structural foam have been written covering processing techniques, properties and machine development. This book differs from them in that it is not a processing manual. It is written primarily for designers, design engineers, and purchasing personnel seeking an answer to the question "shall I design and buy the plastic product I need in *structural foam*?"

Bruce C. Wendle

Contents

STRUCTURAL FOAM

1

What Is Structural Foam?

I. INTRODUCTION

First we must define exactly what structural foam is. This is not an easy task because it is different things to different people. Thus, one must define exactly which structural foam is being discussed.

II. STRUCTURAL FOAM AS A SYNTHETIC MATERIAL FORM

Above all else, structural foam is a form of plastic polymer. Whether molded or extruded, it has all of the glorious advantages of plastics, synthetics, polymers—whatever you wish to call them. It also has the same disadvantages.

A. Advantages

Let us take a look at some of the attributes normally associated with plastics:

1. It is lightweight when compared with metals.
2. It can be colored nearly any color.
3. It can be molded into nearly any shape or form
4. It has many forms with many properties. Over 39 basic families of polymers are used regularly and new ones are being developed daily.

5. It is energy efficient to produce. A graph showing just how efficient is included (see Fig. 1.1).

B. Disadvantages

The disadvantages of plastics are more a result of misuse of the materials. Over the years plastics, because of their low costs, have been used in applications where they have not performed satisfactorily. An example is the toy industry where broken plastic toys were found almost as soon as they were pulled from the box. From these misuses have developed the "cheap" connotation associated with some plastics—even today.

The second problem in the use of plastics is the lack of consistent physical testing data on the various materials. Each material supplier tends to supply only positive information on his materials and, although standard ASTM tests are available for plastic materials,

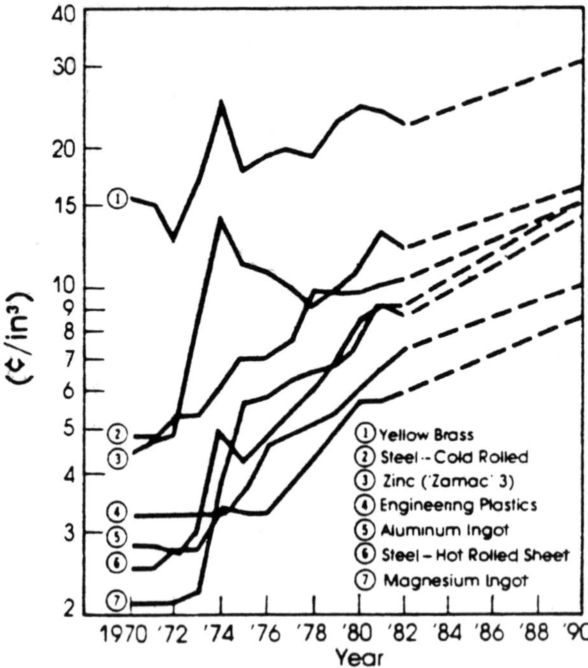

FIGURE 1.1 Energy costs to produce various materials. Source: Plastics Focus, 1984.

changes are often made in test procedures that favor one material over another. The fact that plastics are a relatively new material when compared with metal also limits the amount of consistent data available.

Another factor that has played an important part in some misapplications has been the practice of adding large amounts of regrind or reworked resin to reduce costs. For most applications, this has no apparent effect on properties. However, the material can be definitely changed from what the material supplier provides.

Foams produced during this period were usually accidents caused by excess moisture in a standard injection molding resin. This produced parts with bubbles, an undesirable effect.

C. Plastic Application

Because the technology is changing so fast, it is almost impossible to predict what new applications will be won over by plastics. As new products are produced, each one must be considered a potential plastics application. Obviously all will not be made in plastics. However, the combinations of plastics with other materials such as metal, sand and hundreds of other nonpolymerics will increase the places where it will be utilized. The ability to foam these products is just another means of producing a product to fill a demand.

III. MATERIAL FORM OR PROCESS?

Structural foam means various things to many people. Is it a material form or a process?

A. Society of Plastics Industry Definition

The Structural Foam Division of the Society of Plastics Industry, (SPI), defines it as "a plastic product having a cellular core, integral skins and a high enough strength-to-weight ratio to be classed as structural" [1].

To get a firm understanding of structural foam, let's break down this definition into its various parts.

A plastic product. This can be either thermoplastic or thermoset. It can include filled or unfilled product (glass fibers, carbon fibers, etc.), and can include molded or formed product produced in variety of ways.

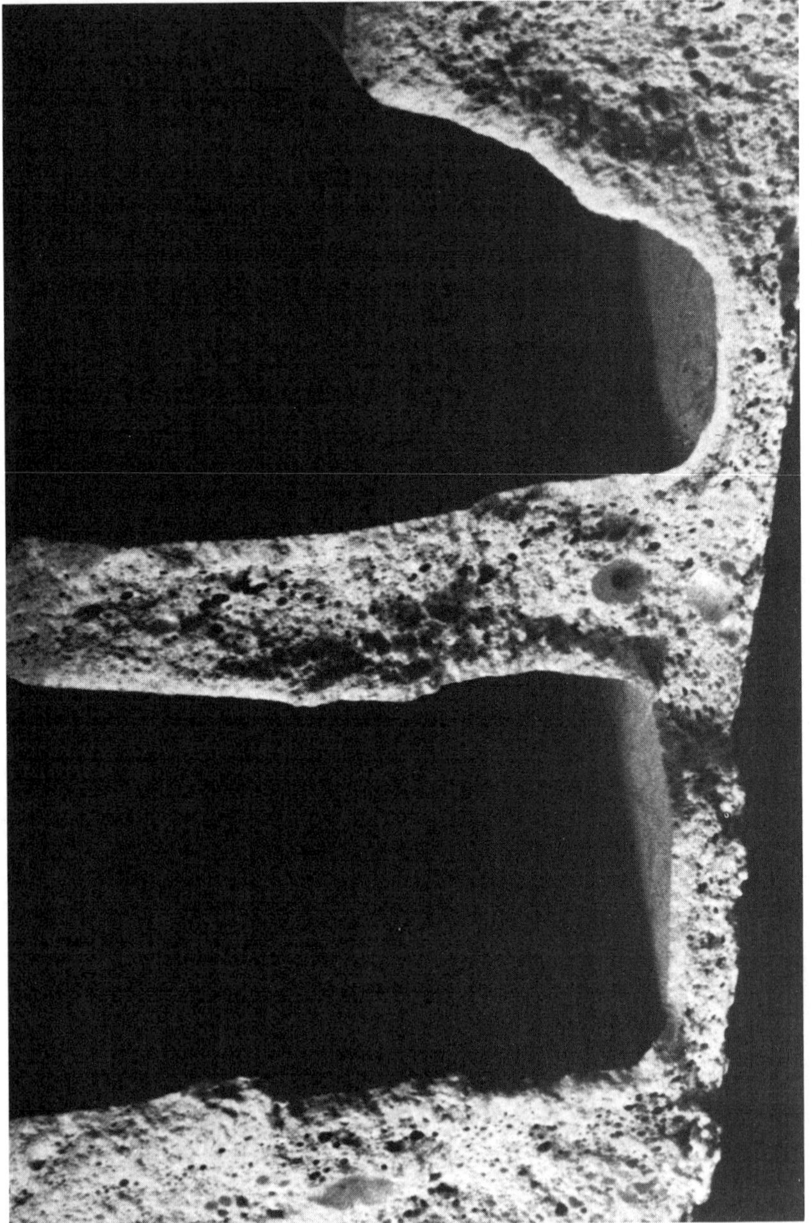

FIGURE 1.2 An example of structural foam showing the integral skin and cellular core.

A cellular core. This generally means a reduction of density in the core or center of the part. These voids can be initiated by a number of things such as nitrogen, ammonia, water vapor, or carbon dioxide but are normally all filled eventually with air.

Integral skins. From the skin formation on the outside of the part comes the strength. This feature is what separates structural foam from other plastic foam products such as expanded polystyrene, or EPS as it is often called.

High enough strength-to-weight ratio. The strength-to-weight ratio is really a relative term coined to compare one material with another. Since the word "strength" is rather nebulous, one must define it as well. Do we mean impact strength? Tensile strength? Compression strength? All are part of the strength of a given material. Certainly the definitive part of the term is weight.

For a load-bearing application, compression strength may be used for practical comparison. Tensile strength would also be useful. Generally plastics have a low tensile strength as compared with other materials.

Structural. This is the catch-all phrase. All products are structural in one sense since they all must at least support their own weight. However, this phrase was added to keep the definition broad enough to include any structural foam without having to apply an exact value to the strength-to-weight ratio.

From the above you can see why we have a very difficult time explaining just what is structural foam and what is not (see Fig. 1.2).

For the purposes of this book we will discuss a number of material forms that fall well within the definition and which should be considered in applications where there is commercial justification for their use.

IV. THE HISTORY OF STRUCTURAL FOAM [2]

How do you begin a history on a material form? Could it be, as my good designer friend Lauri Plaskan says, that "it all started with the animal skeleton." At any rate, the form of a hollow or foamed center core with integral skins has got to be a natural answer to many of our structural problems. As plastic materials go up in cost the idea of using less material to do the same job also becomes more and more enticing.

Since we all agree on that, where did plastic structural foams start? Obviously foam or cellular products find their origin in nature. Sponges, wood, bones—even beehives—are formed of open

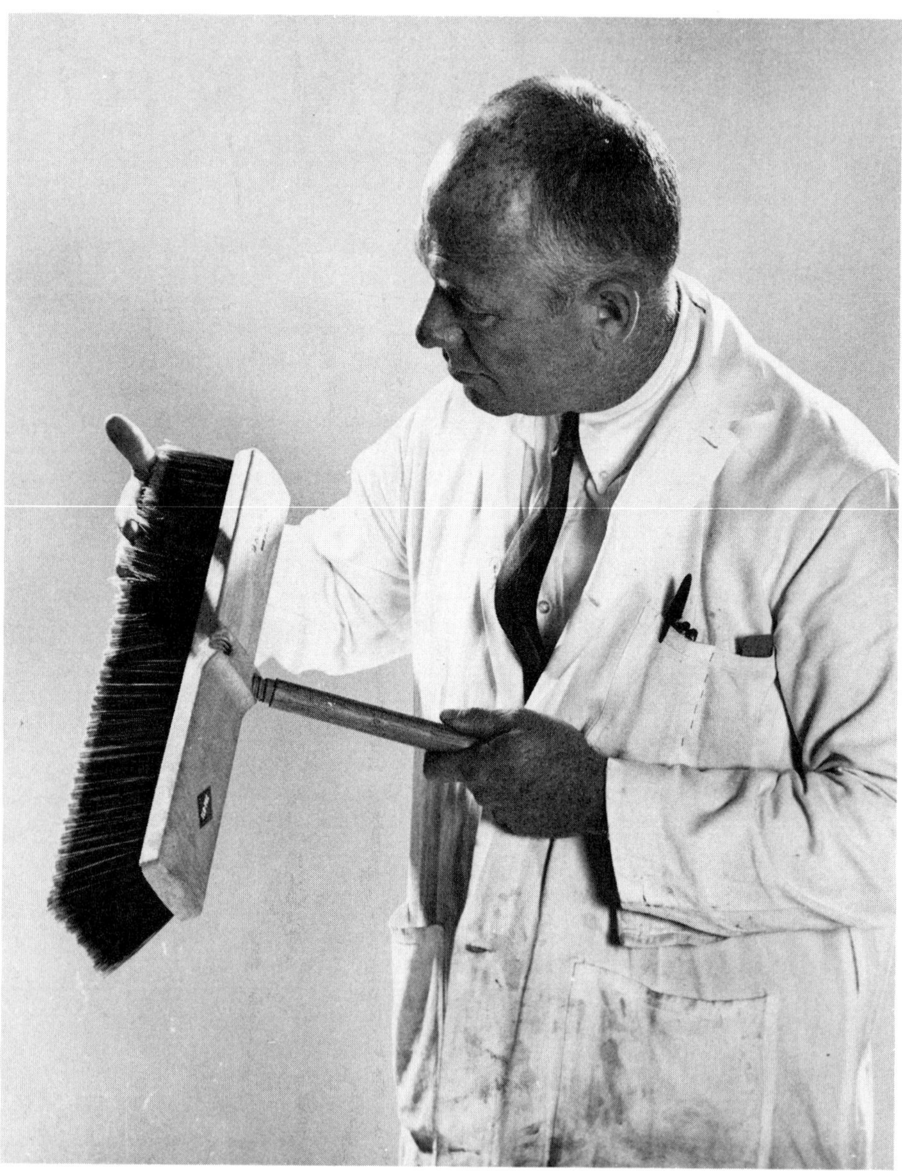

FIGURE 1.3 The first commercial product produced by a custom molder of structural foam. This push-broom block failed because the technology at the time could not guarantee a void free structure. Paint brush handles and other types of brush handles are now being successfully manufactured from foam.

or closed cells. On the basis of this, foams are pretty old stuff. However, when you look at the relatively short time so-called plastic or synthetic materials have been around, plastic foam structures, especially structural ones, do not go back too far.

The first rigid foams, those produced using the wonders of urethane chemistry, were developed in 1937 by a German, Dr. Otto Bayer. These were developed to supply the German war machine with lightweight, high-strength sandwich materials for aircraft, submarines, and tanks.

Then for a number of years, the rigid market sputtered and flittered with not much commercial activity. The flexible foams became the big thing and all kinds of soft cushioning material was produced for mattresses, pillows, and upholstered furniture. Expanded low-density polystyrene foam also became a factor in the packaging area during this time. About the only rigid foams produced during this period were when an injection molder got his resin wet.

It was not until the 1960s before the industry really moved into the so-called structural foam we know today. Richard Angell and his colleagues at Union Carbide's Bound Brook plant had been looking for new ways to make use of some of Carbide's raw materials.

In 1963 they filed a patent on the "method and apparatus for injection molding formed plastic articles." The first licensee signed up by Carbide was Clinton Plastics, Inc., a small custom and proprietary molder in Clinton, Massachusetts. On the basis of an agreement that virtually guaranteed the salability of a foamed push broom block, Clinton purchased a 200-ton Williams-White machine and started out to produce and market this new product (see Fig. 1.3).

The final demise of the product was not because it was not a good idea, but because technology at that point was not capable of producing a block without large voids. These large voids caused great embarrassment when the staple inserted into a drilled hole in the block to hold the bristles failed to find any plastic to hold into.

Not to be completely defeated by this setback, Clinton Plastic's management formed a new company appropriately named First Rigid Foam, Inc., and the first structural foam custom molder was launched. By this time other licensees had been signed and a fledgling industry was formed.

V. THE VALUE OF STRUCTURAL FOAM

Why would a design engineer choose a foamed product in the first place? Why not the tried-and-true method of injection molding a solid, smooth-skinned, conventional product?

FIGURE 1.4 Very large parts can be produced on today's big equipment. This 90 pound cabinet is made on a 500 ton structural foam machine using machined aluminum molds.

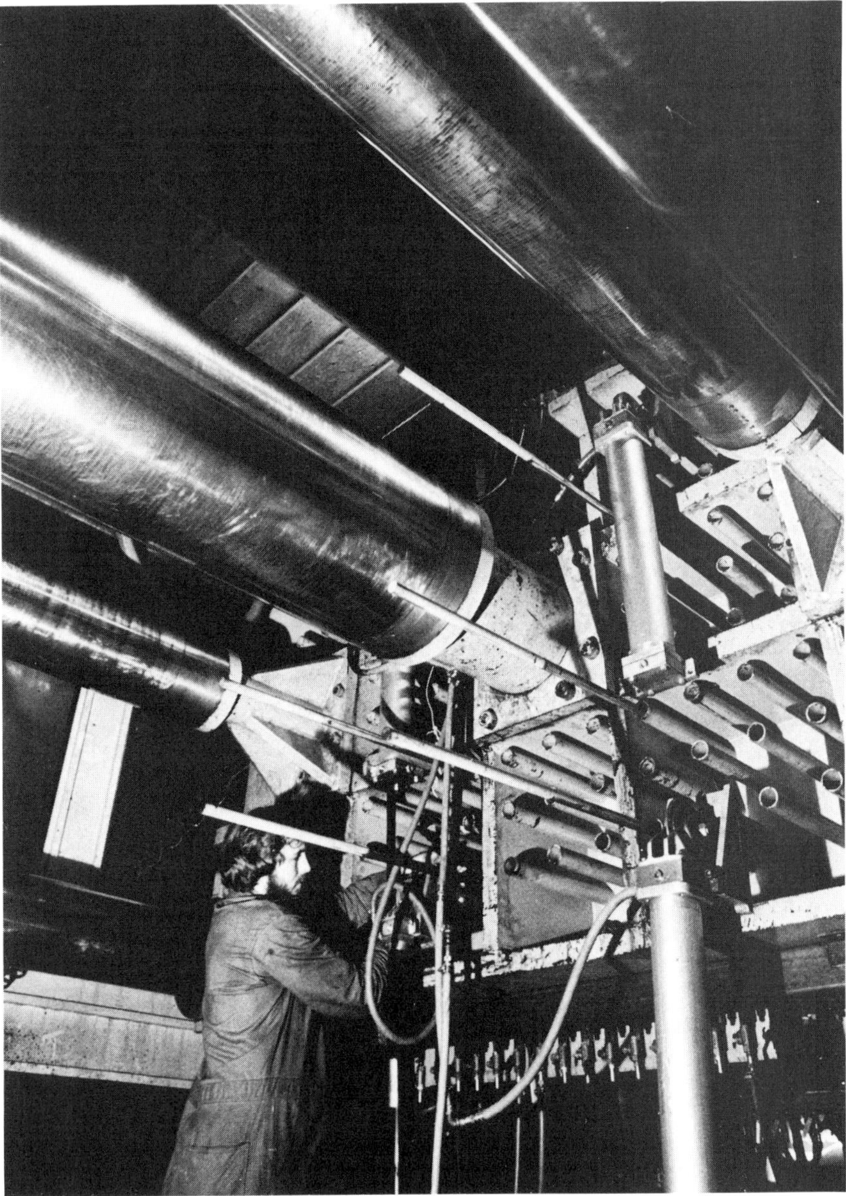

FIGURE 1.5 From the inside of a 500-ton structural foam machine one can get an idea of the size. This horizontal machine has the capability of shooting over 100 pounds of thermoplastic in one shot.

The obvious answer of weight reduction due to the cellular core, surprisingly, is not the main reason and, in fact, may not even be an accurate answer. Although the core density is lower, the need to go to thicker cross sections and the relative low-density reductions used make weight reduction a questionable advantage.

The true value of foam lies in the increase in strength-to-weight ratio and in the reduction of sink marks in areas opposite thick sections. Other values include a significant reduction in tooling costs due to the lower pressures encountered in the cavity during most types of processing.

The structural integrity of a properly formed part is also greatly improved due to lowering of stresses while the part is being molded, as well as a general increase in stiffness due to an optimum wall thickness of one-quarter inch. Higher-pressure foams with thinner walls retain some of these properties and also provide improved surface conditions.

The main disadvantage of most structural foams remains the need to finish the product to achieve an acceptable appearance for such products as business machines. Much work is being done in this area, and future types of structural foam may not require this expensive step.

The ability to produce large parts (120 lb) or sets of parts on one machine remains the biggest single advantage of multinozzle equipment, whereas medium to large parts can be produced faster and with less surface swirl on the newer, single-nozzle machines (see Fig. 1.4). Many foam applications are being molded on conventional injection equipment with excellent results.

The biggest problem with large parts is the need for large molds. The cost of a large tool jumps dramatically for shots above 15-20 lb. Tooling material is very expensive and, even at the low pressures of structural foam, the mold must be built to retain a fair amount of pressure. However, for producing a molded plastic part of a large size where all wall sections are controlled and resonable dimensional tolerances can be held, structural foam is an excellent candidate once the tooling expense is overcome (see Fig. 1.5).

According to foam pioneer Jim Hendry, "for the capital dollar invested, structural foam injection molding machine (properly loaded) gives you more shippable pounds per hour than any equally priced injection molding machine" [3].

VI. THERMOPLASTIC STRUCTURAL FOAM APPLICATIONS

Who's using it and why? Further discussion into just where foam is going may help the designer decide whether the material fits his needs. A very reliable guide for the use of any material form is the application of the material to various design areas. Structural foam first found acceptance as a replacement for wood in furniture parts. Only limited success was found in this area due to changing styles and relative short product runs. However, thermoset foams and some styrene foams are still used in this market.

A. Instrument Housings

The largest and still growing area for foam is still enclosures for the computer peripheral field. High-speed line printers, CRT terminals, and hundreds of other instrument housings are being produced in the various engineering or high-tolerance materials such as polycarbonate, ABS, flame retardant polystyrene, and modified polyphenylene oxide. Complete structural foamed desk top work stations are in use and multipart copier housings are coming down the production line (see Fig. 1.6).

B. Material Handling Markets

Next to housings for the computer peripheral market, material-handling applications of all kinds are probably the biggest user of structural foam. An example is shown in Fig. 1.7. Large bins carrying 1000 lb and having four-way pallet entry are seeing wide use. This high-density polyethylene resists corrosion, ultraviolet attack, and ambient temperatures. Tote bins, buckets for picking fruit, nuts, and vegetables, and trays for drying are all established applications. Most are in polypropylene or polyethylene, depending on temperature requirements. Platform trucks are being molded in high-density polyethylene.

C. Construction

Structural foamed materials, especially high-impact styrene, polypropylene and high-density polyethylene, are being used in such

FIGURE 1.6 Many applications in the information processing, transportation, appliance, and material handling industries which were once metal have found rapid acceptance in structural foamed materials.

FIGURE 1.7 Large bins molded of structural formed high-density polyethylene have found wide acceptance in the fishing industry as well as in the plastics industry itself where both finished goods and molding resin are stored.

items as door sashes, window frames, and even doors themselves. Because large molded parts up to 100 lb can be produced at a rapid rate, other applications will be forthcoming. Problems that must be overcome include combustability requirements in building codes. The ability to mold special effects such as wood grain (see Fig. 1.8) or brick front could be a factor in someday using the material for panels. In the institutional market, cabinets of structural foam are finding good acceptance. Drawers of structural foamed polypropylene have been available since 1971.

D. Furniture

Beside the items mentioned above, structural foamed high-impact styrene has been used in chair frames and other similar nonappearance structural parts. Wood-grained parts were some of the first applications as far back as 1968. Figure 1.8 is a good example.

FIGURE 1.8 Fine example of a structural foam panel molded in an oak wood-grained effect for use on a stereo. Several different styles are molded using cast aluminum inserts in a machined aluminum chase.

Fold-up bathroom vanities have been on the market as well as complete kitchen cabinets and doors. Here again, the ability to mold large parts having good structural integrity has been the reason for use.

E. Leisure Market

The leisure market includes such foam applications as boat seats, swing platforms and a wide range of toys and sporting items. The ability to mold a variety of parts at one time has been a big economic advantage. Resistance of these parts to outdoor weathering is improved by the low stress levels found in structural foam. A swim platform is shown in Fig. 1.9

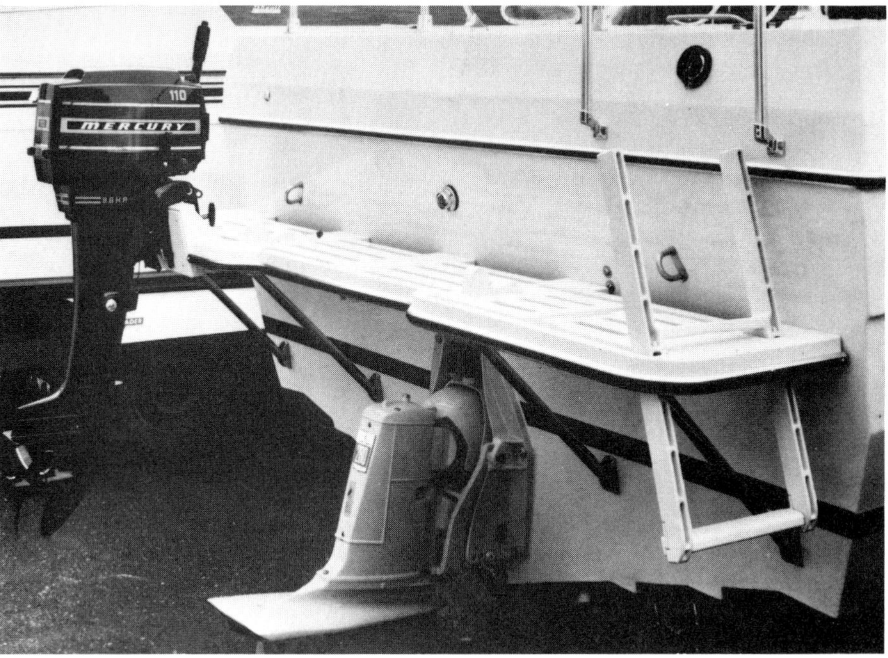

FIGURE 1.9 One piece, 90-inch swim platform for boats is produced in Seattle by the Tempress Company. Material is UV-inhibited high-density polyethylene with 5% glass fiber added for rigidity.

F. Automotive and Farm Equipment

A great future is expected for structural foam in automotive application. Tractor tops and trunk instrument clusters are a few of the applications coming off the boards. Several automotive companies have put in big molding machines, and the amount of foamed plastic on the average vehicle is bound to increase.

G. Specialty Markets

Specialty markets for such things as foamed sheet or extrusions have really not developed to any great extent. The insulative characteristics as well as the structural integrity of such a material might be useful. Some use of extruded profiles for framing, edging, or moldings has been found. Foamed vinyl moldings are produced in some quantity in Canada.

H. Packaging

Packaging applications are limited due to high cost of tooling. Some use has been made of structural foam where boxes for instruments or weapons require complex rigid internal shapes. Most of this type of packaging is done with expanded polystyrene and not structural foam.

I. Future Uses

Several areas of application seem to be naturals. Flotation units such as floats, bouys, and pontoons would seem likely candidates for polyethylene foam, and small boat hulls are a possibility once bigger machines become available. Several internal parts on large trucks could be made of structural foam, saving both dollars and weight. Large dish antennas for TV reception systems are now constructed of foam.

According to Richard G. Denton of Midland Ross Corp., "almost all of our structural foam products are replacements for traditionally metal products" [4].

New technology is bringing about many potential changes, and the industry must be prepared to grow with them. The old ways of molding a cellular product may not be the best way. Application areas once thought to be exclusively the right place for low-pressure thermoplastic foams may be challenged by newer materials and processes.

VII. THERMOSET STRUCTURAL FOAM APPLICATIONS [5]

The versatility inherent in polyurethane chemistry and the advantages of the reaction injection molding (RIM) process make possible the utilization of polyurethane structural foams in most of the applications being considered for structural foams.

A. Reaction Injection Molding (RIM)

Other applications use the unique properties inherent only to RIM polyurethanes. For example, RIM polyurethanes enter the mold as a liquid, thus permitting the encapsulation of various inserts such as wood, metal, and fiberglass.

B. Applications Using RIM

Water skis, for instance, are a composite consisting of polyurethane structural foam that bonds the top covering and encapsulates steel rods, thus providing the correct combination of flex and structure. Office chair arms and bases have a more "friendly" appeal by molding a RIM integral skin foam around their metal parts. This composite technology is also utilized in the construction industry where skylights, window frames, and solar collectors are manufactured by encapsulating wood and metal extrusions with RIM structural foam. The material is dimensionally stable in humid conditions and retains its properties in outdoor exposure.

A stainless steel liner is totally encapsulated with polyurethane structural foam in the production of Miller Brewing's "poly keg" (see Fig. 1.10). The urethane jacket protects the container from damage. The design versatility allows for molded-in handles and rolling rings, which facilitate carrying and stacking of the kegs.

RIM polyurethanes can easily flow behind various surfaces to produce unique composites. Luggage for instance, is manufactured by injecting a RIM structural foam behind a sheet of vinyl that has been vacuum formed. Similarily, large viewing screens for home video sets are produced by injecting structural foam behind the aluminum projection surface of the screen.

RIM polyurethanes can also be molded economically into large parts. The Jimmy Connors Rally Champion Tennis background consists of a number of panels bolted together to form a 14 ft × 18 ft surface.

FIGURE 1.10 RIM molded rigid urethane structural foam is used around this aluminum beer barrel. Handles and stacking ribs are some of the molded-in features. The insulative qualities of the foam also add to the application.

The agricultural industry utilizes RIM polyurethanes for tractor body parts such as radiator grille frames, fenders, and cab roofs. Another agricultural part is the hay conditioner roller, which consists of 60 lb of urethane molded around a steel core 109 in. in length. These large parts are not only produced economically with RIM polyurethanes, but they also have excellent corrosion resistance and durability, even with constant outdoor exposure and rough handling.

The swirl-free molded surface of RIM structural foams provide an advantage to the appliance industry for air conditioning housings, freezer doors, and evaporative cooler housings where cosmetic appearance is important. Herman Miller's Burdick group, for instance, is a modular system for office furniture consisting of surfaces and units supported on a beam. Other decorative furniture with a woodlike appearance is molded by merely texturing the mold surface.

VIII. FUTURE OF STRUCTURAL FOAMS

It is hard to imagine any limitation on the future of structural foam, assuming the continual development of larger and larger machinery and more sophisticated tooling techniques. Applications such as the huge 100-lb vegetable tote bin and a one-piece wall cabinet system seem to indicate there is no limit to how large a part can be produced using a structural foam processing technique. Shrinkage problems are of course, going to be a continuing problem as the parts get larger and larger. However, mold design techniques should be developed to take care of this problem. It is not inconceivable to think of parts weighing up to 300 or 400 lb being produced on large presses and using such new material developments as glass or carbon fiber-filled polymers giving extremely high physical properties at relatively low material costs.

The ability to produce a complete set of parts on one machine for a given assembly is also a valuable tool and allows the designer more latitude when designing parts that need to go together to form complex assemblies. Various fillers used in these structural foam materials not only provide the ability to change the physical characteristics of the material, but they also act as nucleating agents that help to produce more uniform cell structure throughout the part. Although the use of chemical blowing agents to produce structural foam is just one more technique to use existing plastic machinery, it is obviously going to become more and more accepted as a way of producing the polymeric products demanded by the customer.

It is reasonable to assume that the utilization of this material form has only just been scratched and that larger volumes of structural foam will definitely be utilized in the future.

A. Market Growth

According to Michael Colangelo, associate editor for *Plastics Technology* magazine,

> "More than ever, the state of the structural foam market depends on technology developments aimed at making the process more economically viable.
>
> At present, the majority of structural foam processing is handled by about 125 molders, who processed an estimated 179.6 million pounds of resin in 1982.

1. Cabinetry

Consumer	Electronic	Industrial
TV cabinets	Printers	Equipment housings
Phonographs	CRT terminals	Machinery enclosures
Speakers	Disc drives	Military enclosures
Clocks	Medical instruments	Telephone cable enclosures
Drawers	Game housings	
Drawer fronts	Copiers	
Humidifier cabinets	Bank teller machines	
Air coolers		

2. Material Handling

Pallets	Trays
Milk cases	Picking crates
Soda cases	Shipping containers
Drums	Dunnage
Bins	Chicken coops

3. Industrial

Battery cases	Mop buckets
Transformer housings	Wringers
Trash containers	Carts
Vehicle crash barriers	

4. Consumer

Recreational	Musical	Miscellaneous
Boat ladders	Organs	Toys
Swim platforms	Pianos	Childrens' furniture
Boat seats	Guitars	Toy chests
Diving platforms	Instrument cases	Gun stocks
Oars and Paddles		Bird baths
Dinghies		Bird houses
Ping pong tables and paddles		Planters
Portable camping kitchen		Garden fencing
Toboggan		Utility tubs
Sled		Buckets
Game tables		Coffins and vaults
		Urns

5. Construction

Shutters
Shingles
Bow window frames
Doors, interior
Storm doors
Kitchen cabinets
Bifold doors
Underground junction boxes

6. Furniture

Decorative panels
Interior paneling
Valence and trim
Chair backs
Chair shells
Chair pedestals
Chair and sofa arms
Bed frames
Head boards
Drawers
Picture frames
Hospital handling system

7. Automotive and Aircraft

Glove box
Door sills
Bench seats
Bucket seats
Fan shrouds
Fender liners
Instrument panels
Window trim
Vehicle tops—Agricultural
Station wagon floor

8. Miscellaneous

Bicycle frames
Antennas
Traffic signal housings

Examples of some of these applications are shown in Figs. 1.12, 1.13, 1.14, 1.15, 1.16, 1.17, 1.18, 1.19, and 1.20.

C. Future Applications

In 1979 Russel Kidder [8] in a talk before the seventh SPI Foam Conference in Norfolk, Virginia, said, "it [the structural foam industry] is a very fragmented industry covering many different end uses and utilizing many different resins in different types of machinery."

Today, 5 years later, the industry is still fragmented with new types of machinery, new resins, and many new market uses.

A list of future applications will certainly include:

1. Low-level radiation waste containers
2. Canoe and dinghy hulls

FIGURE 1.12 A large structural foam plant manufacturing large panels for the swimming pool industry. A large area is needed to store both the raw material and finished goods for such a facility.

3. Computer furniture
4. Military electronic housings (radios, etc.)
5. Robot components
6. Electronic game housings
7. Decorative roofing material
8. Aircraft components
9. File and storage cabinets
10. Shipping containers

The wide range of possibilities is apparent in this list.

This wide application base may, in fact, be the strength of the industry. If, in fact, the market for electronic housings in structural foam is limited, new market areas will most certainly take up the slack.

FIGURE 1.13 A sun dial produced from structural foam is one of the many consumer products made today.

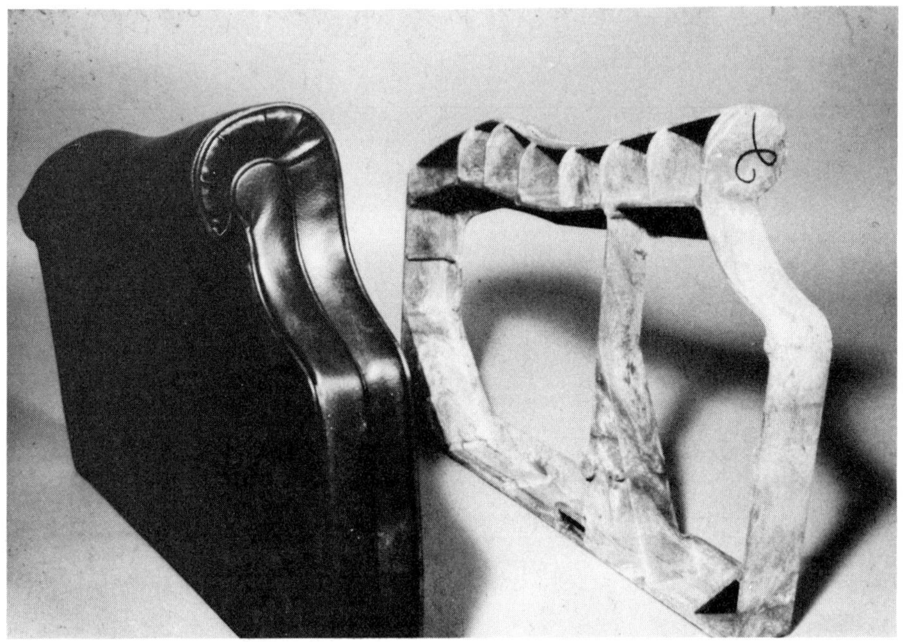

FIGURE 1.14 Example of a structural foam chair frame and the finished product.

D. Forecast of Growth

As put together by Predicasts, the Cleveland, Ohio, market research firm, for its new study on Plastics Foam Markets, the figures indicate a 77% increase from 3.5 billion lb in 1982 to a projected 6.1 billion lb in 1995. Urethane foams will remain the leader (with structural applications in RIM starting to come on strong during the 1980s); styrene foams will stay second [9].

FIGURE 1.15 A bi-fold door produced of high-impact polystyrene structural foam being removed from a multinozzle machine.

FIGURE 1.16 This popular picking bucket is produced from foamed high-density polyethylene.

FIGURE 1.17 The advent of the drive-up banking window has also opened up new applications for structural foam. This large housing is typical of the products being designed in this field.

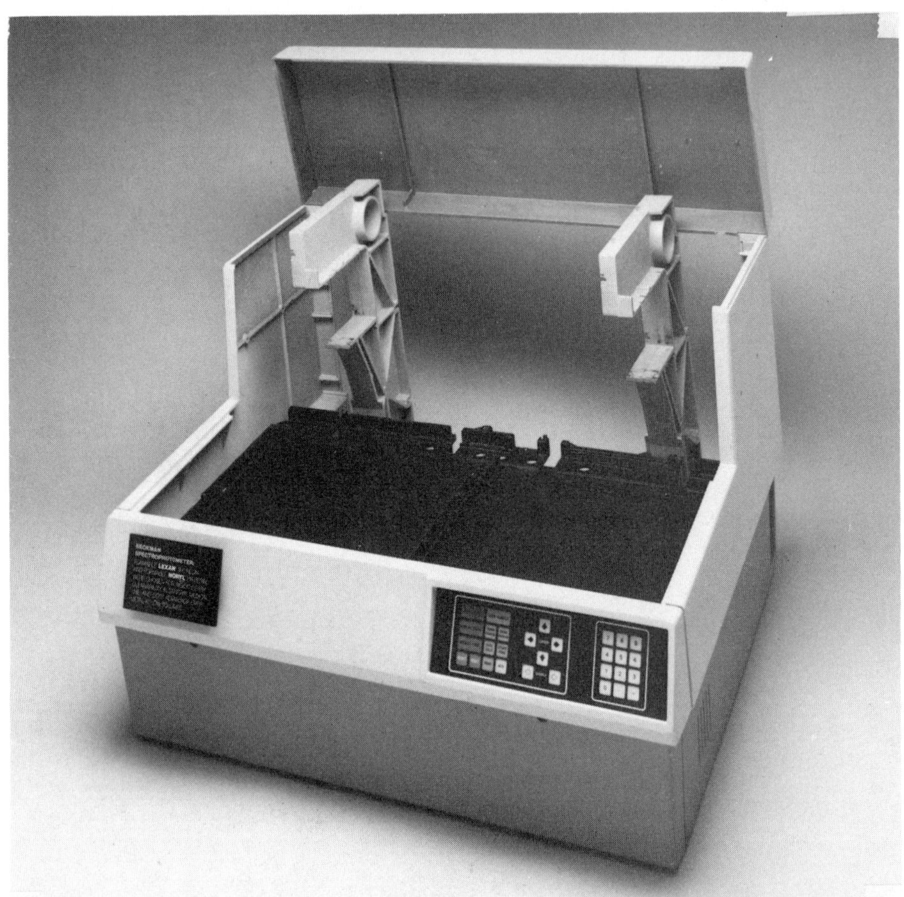

FIGURE 1.18 The medical electronic revolution has brought about many new structural foam applications. This liquid scintillation machine produced by Beckman Instruments, Inc., uses a combination of different structural foam resins to provide a substantial cost savings over metal. The polycarbonate chassis supports over 300 pounds.

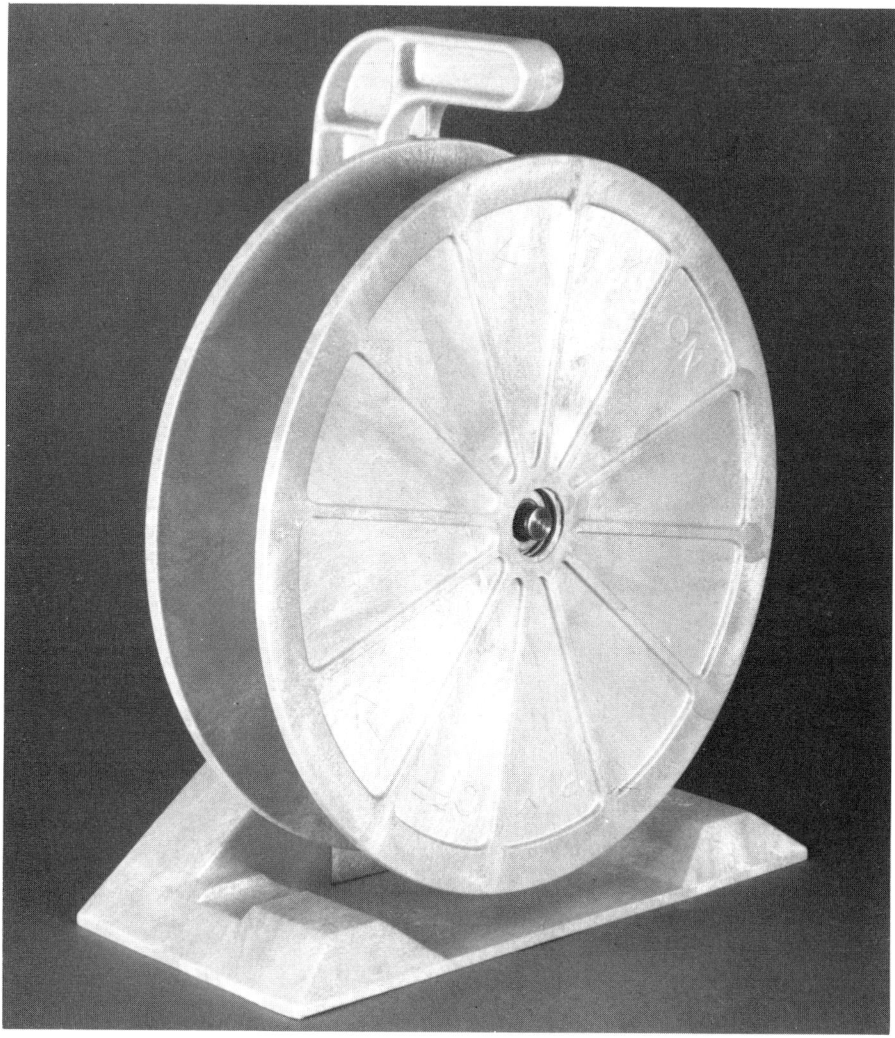

FIGURE 1.19 Telephone wire reels manufactured from polycarbonate structural foam are lighter in weight and less expensive to produce.

FIGURE 1.20 This high-speed line printer produced by Data Products of Woodland Hills California, utilizes modified P.P.O. structural foam.

IX. STRUCTURAL FOAMS BUT NOT "STRUCTURAL FOAMS"

Because a cellular core is not a unique material form, it may be helpful to discuss other plastic material forms, i.e., those material forms that are cellular but not necessarily classed as structural foams.

A. Foamed Sheet Stock

Several sheet stock suppliers have developed materials that have a solid skin and a cellular core layer and can be formed much like solid sheet. ABS and vinyl sheet both are available with a foamed core. Rigid urethane foam has also been used as a backing material for ABS or vinyl and forming, while limited, can be accomplished.

The properties of these composites are an improvement on the materials individually.

Insulation. Both heat and sound insulation characteristics are improved.

Stiffness to weight. In some composites, the use of foamed backing will provide much higher stiffness without adding substantial weight.

Applications. Most uses for these sheet stock combinations are found in the construction industry. However, small-volume, vacuum-formed products can be produced.

B. Expanded Polystyrene (EPS)

Polystyrene, blown with pentane gas, has been accepted as a packaging media for many years. Density can be controlled fairly well and some functional parts other than packaging have been molded. Normal density for packaging foam is 1 1/2 to 3 lb/ft^2. Higher densities up to 30 lb/ft^2 have been produced for special applications.

Processing of this product employs a pre-expanded stage where steam is blown through the molding bead prior to putting it into the mold. Before molding the bead is put into large net bags to dry. After drying, the material is then blown into a mold (usually cast aluminum) and treated again with steam for final fusing. Finally, it is cooled with cooling water and ejected from the tool. The molded material does not develop a skin and is basically made of a system of fused cells.

The largest segment of the EPS market in annual consumption is the block-molding business. Large blocks of the material are produced and cut into sheet stock, which finds much use as insulation for the construction industry. These blocks are also used for flotation.

The energy-absorption characteristics and heat and sound insulation are the best properties of EPS.

C. Self-Skinning Urethane

Another foam system hard to class as structural is the flexible self-skinning urethane foam. Available in both firm and soft forms, the material is the product of a reaction between a polyol and an isocyanate combined with a blowing agent. The product in the molded form has a tough skin of polymer surrounding a cellular core.

Most applications are found in the automotive industry where arm rests, seat cushions, sun visors, and instrument clusters have been produced from this material for many years. Recent developments have combined a self-skinning urethane over a frame of thermoplastic structural foam to produce an exceptionally tough, good-looking dashboard for a large truck manufacturer.

D. Extruded Thermoplastic Foams

Improved economics in the construction industry have opened up a rapidly expanding market for another foam form. Extruded profiles of foamed polyvinylchloride (PVC), polystyrene, ABS, and polyethylene are being produced for use as trim and moldings that are replacing the conventional wood products.

Low-density polystyrene in a thin foamed sheet form is used in large quantities for egg cartons and snack food containers. This product is usually extruded in special extruders and then formed to the required shapes. Covers for glass bottles, a proprietary development by Owens-Illinois, features foamed material for insulation.

Pipe extruded in foamed PVC or ABS is being produced and marketed by Armotech Corp. HPM Corporation has developed a coextrusion technique for producing a cellular ABS pipe with a solid skin.

E. Syntactic Foams

This very special type of foam features small preformed bubbles of glass, plastic, or ceramic mixed into a thermosetting resins such as epoxy. Somewhat heavier and stiffer than gas-blown foams, these composites have excellent thermal and acoustical properties. Sometimes combinations of both gas-blown and preformed bubbles are used. Polyester is also used for a binder in the stiffer foams, whereas a more flexible system can be produced by using a more linear polyurethane resin.

Application areas for syntactic foams include the electrical and electronic industries, aircraft, space, and the military.

F. Summary

All of the above materials, when used in certain applications, can be classed as structural, and all are most certainly cellular in nature.

REFERENCES

1. *Structural Foam*, Society of Plastics Industries, 1982, p. 20.
2. Bruce C. Wendle, *History of Structural Foam*, 1976 Conference Procedings, S.P.I Structural Foam Division, Technomic Publishing Co., 1976, p. 12.
3. James Hendry, *Structural Foam Processing and Application—Yesterday, Today and Tomorrow*, 1978 Conference Proceedings, S.P.I Structural Foam Division, Technomic Publishing Co., 1978, p. 31.
4. Richard G. Denton, *Use of Structural Foam in Commercial Products*, 1980 Conference Proceedings, S.P.I. Structural Foam Division, Technomic Publishing Co., 1980, p. 99.
5. *Structural Foam*, S.P.I. Structural Foam Division, 1984.
6. Michael Colangelo, Structural Foam: Where It's Headed, *Plastics Technology*, June 1983, p. 41.
7. Michael Colangelo, Structural Foam: Where It's Headed, *Plastics Technology*, June 1983, p. 41.
8. Russel Kidder, *Structural Foam Futures—An Overview*, 1979 Conference Proceedings, S.P.I. Structural Foam Division, Technomic Publishing Co., 1979, p. 56.
9. Joel Frados, *Plastics Focus*, Plastics Focus Publishing, Inc., 15:2 No. 29, Sept. 12, 1983.

2

How Can You Utilize Structural Foam?

I. INTRODUCTION

For several years now, structural foam has meant different things to different people. As a material form it provides a product with a solid structural feel, little if any warping, and, once painted, an excellent-looking stable product. It has also been expensive when compared with injection-molded parts requiring no finishing.

The big advantage really has been tooling costs. When injection-molded parts are quite large, the high-pressure steel tooling needed is very, very expensive. At this time, the two factors that most influence the swing to or away from structural foam are still part size and tooling costs. with electronic packages getting smaller, there has been a definite swing toward smaller housings and, with it, a swing back to injection molding. For those looking for other product features such as sound deadening, heat insulation, and a solid, weight-supporting structure, structural foam is still a winner.

We should also point out that the surfaces of most injection-molded thermoplastics are soft and can be scratched rather easily. Thus, for a tough, scratch-resistant, chemical-resistant surface on a solid molded part, it is hard to beat a two-component polyurethane coating on top of a structural foam-molded enclosure.

If a product requires a large industrial part that is molded in one shot and has a great deal of integrity, stiffness, and toughness, a structural foam molded part would be a likely candidate.

When evaluating the effect of tooling cost, the volume expected to be run is the dictating factor. With a housing of which only 1000—3000 are required per year, it is difficult to justify a $40,000 tool. If a machined aluminum mold can be built for $15,000–$20,000, a more expensive piece price can be justified, especially if it does a quality job.

II. ENGINEERING AND DESIGN DECISIONS

Whatever the product, all designers, design engineers, and product design people are faced with the same problems:

1. Picking the right material and process
2. Developing it on a tight time schedule
3. Keeping the cost within budget

Structural foam can provide an answer to some of these problems. In an article in *Plastic World*, Three designers faced with the above problems had this to say:

Nami Hazneci, manager of the Burroughs Corp. plastics and rubber laboratory noted, "If a material isn't going to survive in the field, then there's no use molding the part" [1].

With regard to volume considerations, "In a desk top housing, injection molding will always be less expensive than structural foam . . . if quantities are about 5000 units annually," says Hazneci. "Projected volumes two and three years down the road must be considered," he continued.

Where the parts are larger, say for floor units, foam might be considered because of the size of the part alone.

In a case involving a housing for a terminal for Hostech, Inc., the world's largest supplier of graphics and editing equipment for newspapers, James Gallagher, manager of engineering services, indicated, "tooling costs were the deciding factor." The project in this case went to rigid urethane foam molded or a RIM machine. "A housing's success depends on good mechanical design and close work with your molder," he noted [2].

In the third case history, Jerry Nunn, vice president of operations at Applied Digital Systems (ADDS) made these comments about the choice of structural foam for housings used on three lines of desk-top computers produced by the company; "volumes and tooling costs are the most critical considerations during process selection" [3].

Generally, tooling for structural foam costs is at least 30% less than for injection-molded housing, especially in the larger parts. The varies dramatically with the part complexity and with the type of equipment that will be used for molding the part. This differential will decrease as higher pressures are used to improve the surface.

Designers all have the same initial question: When do I use structural foam and when do I use some other form?

III. MATERIAL AND PROCESS SELECTION

If one can believe the literature sent out by material suppliers, there is only one material to use for every application—theirs! Before you can possibly select a material, or a process you must know your product.

If there is one place that plastic users waste both time and money, it is in premature design. Decisions made because "Joe Engineer" knows the resin salesman or because he designed a part that way 2 years ago, and before knowing all the facts about a given product, can cost a company plenty.

A product check list, made before any design work is started, is an absolute must. Such product information plus volume projections, financial conditions for tooling investment, and timing are all important to the decision.

Let us first look at the list of questions one must ask before the design phase can begin. Some answers are easy, others more difficult. Do not be afraid to spend time digging for the information. The right answers now, can save many hours later on.

A. Preliminary Product Review

Before any decision can be made, you must know your product. The following is just one form of questionnaire that could be used to define clearly the parameters one must work with.

PRELIMINARY

PRODUCT DESIGN INFORMATION

Project name (example, printer housing)

General definition of what the product under consideration will do (example, enclose electronic printer mechanism)

What is product environment (example, office or computer room)

Temperature use range ____________________

Exterior temperature range ____________________

Internal temperature range ____________________

Compatability: (What chemicals or solvents might product come in contact with?) ____________________

What is weight or load product must be designed to hold? ________

What are approximate outside dimensions of product? __________

What codes or specifications must product meet (example, combustability, U.L. Laboratories UL-94-V-0)

What volume of this product do you expect to produce over the first year? ____________________

After first year? ____________________

____

B. Seven Basic Questions

There will be many other questions concerned with both function and aesthetics, but these should not be addressed before narrowing the approach. The seven basic questions honestly answered will give your project the direction it needs. One should also continue to ask these questions during the design phase because they can and often do change.

The "Basic Seven" again are:

1. Environment
2. Compatability
3. Size
4. Weight
5. Temperature
6. Codes and specifications
7. Volume

C. Narrowing the Field

From the above information, we can list our options. Do not try to select an approach before obtaining answers to all of the basics. If you do, you may develop some preconceived ideas that will limit your options and cause you to miss the right choice.

The basic material choices are foam plastic, solid plastic, metal, wood, or other. The basic process choices are molded, formed, blown, fabricated, extruded, machined, and cast. Because our main topic of interest is structural foam, our goal is to position or compare it with competitive materials and processes.

For the purposes of this book, let us say that we are able to decide that our product, whatever it is, will be considered only in plastic or metal. Obviously there are other materials of construction that must be considered for specific products; for example, if a pallet is being designed, wood would be a strong contender.

D. Plastic versus Metal

On the basis of our need to decide between plastics or metal, let us review the "basic seven" and how each affects our choice.

1. Environment

Indoor versus outdoor use will determine if a painted surface is required. In plastics, an outdoor application will require a UV inhibitor or a UV-resistant color such as black. A salt water environment would be hard on most metal applications. (See App., Sec. VII.)

2. Use Temperature

Here we tend to separate the materials; generally, the temperature ranges would be as follows:

Plastics—30°F to 375°F
Metals—50°F to 1000°F

In the very low cryogenic ranges and in very high-temperature ranges, there is a special set of circumstances requiring carefully selected materials not normally used in conventional applications. An example would be the filled materials finding their way into engine blocks.

3. Compatability

All applications come in contact with various chemicals, solvents, or compounds during their use life. Some metals and nearly all plastics can be attacked by one or more of these materials. Upfront information will assist in picking the material of construction most likely to resist. In plastics, some processes are more likely to increase stress levels in the part which, in turn, reduces the resistance of the part to specific chemicals. (See App., Sec. VI.)

4. Size

The overall size of parts to be produced will often eliminate certain processes simply due to equipment limitations or to cost.

5. Weight

Approximate weights of both the part and the load to be carried by the part are very critical to the early decision.

6. Anticipated Volume

Probably the most critical of all assessments affecting a project is anticipated volume; not only for the life of the project, but for first-year requirements as well. Accurate market forecasts are difficult to obtain, usually overprojected, and change almost daily. On this shaky rock, one must base a decision.

What normally happens is a "Catch 22" decision with regards to setting economic limitations before selecting process and material—or must one select process and material before knowing the full cost limitation of the program?

7. Codes

Codes and requirements of foreign countries where the product is to be sold will often affect the choices available as well. Each industry is faced with its own range of codes and specifications, and these can have disasterous effects on a project if not considered very early in the selection process. (See App., Sec. VI.)

Codes that most often affect the choice of process and material include:

1. Underwriter labs
2. CSA (Canadian)
3. Federal Communication Commission
4. Federal Aviation Agency
5. Department of Agriculture
6. Federal Drug Administration

E. General Review of Choices

A general review of our choices are given below:

	Metal	Solid Plastic	Foamed Plastic	Wood
Molded	Die cast	Injection molded	Structural foam	Limited
Formed	Sheet	Vacuum formed	Limited	Sheet stock
Blown	NA	Bottles	NA	NA
Fabricated	Available	Available	Not practical	Available
Extruded	Extrusions	Extrusions	Extrusions	Routed profiles
Machined	Available	Size limitations	Not practical	Available
Cast	Casting	Casting	Casting	NA

Variations of all these are available, however, the above methods are most commonly associated as presented.

F. Economics

It is generally felt that one should have a basic knowledge of what affects economics before selecting material and process. Then the top contenders can be selected and a more detailed evaluation of each made. As a general rule, tooling costs go up as piece prices come down. A few parts can be produced by machining or fabrication with little if any tooling. However, costs are high and production rates slow.

Although this relationship should be obvious, it is amazing how many people forget this. It also seems to hold true for both plastic and metal, and one can also look at this with regards to labor involvment. The more that has to be done to the part by human labor; the more expensive, the more quality problems and the slower production will be. Again obvious, but often forgotten.

Automation, including use of computer and robotics, has only enforced these old production rules.

1. Cost Review

For the purpose of this discussion, consider a 4-lb structural foam housing versus a 2 1/2-lb injection-molded part. Both parts have the same square surface area, but the structural foam part has a 1/4-in. wall, the injection part 1/8 in.

Cost evaluation with all other costs being *equal* would be as follows:

Material. Both parts are considered in polyphenylene oxide (injection grade are more expensive)

Foam	4 lb @ \$1.30 = \$5.20	Dif. \$1.13
Injection	2 1/2 lb @ \$1.63 = \$4.07	

Machine Costs

Foam	20 pph × \$50/hr = \$2.50	Dif. \$1.88
Injection	80 pph × \$50/hr = \$ .62	

Finishing

Foam	3 1/2 ft^2 × 1.50 = \$5.20	Dif. \$5.20
Injection	None	

Difference between structural foam part and injection part — \$8.21

Mold Cost

Foam	\$50,000	Dif. \$25,000
Injection	\$75,000 (including surface embossing)	

$$\frac{25{,}000}{8.21} = 3045 \text{ units}$$

At 3000 units of volume or above, it would be worthwhile to look at injection molding. Although properties of these two materials will vary, both may perform the same function. Example: Cover for CRT Terminal.

There are other factors that should also be considered. A few include:

Conductive coating costs
Weight
Freight charges
Stability of housing
Number of separate parts/set
Surface resistance to environment
Future usage of housing

All of these should be considered when making the decision on foam versus injection molding.

IV. STRESS—OR LACK OF IT

The fact that structural foam parts have less stress molded in than injection-molded parts is often mentioned in structural foam literature. However, no further explanation is usually given.

In injection molding, where pressure of up to 20,000 psi are often used to fill and hold a part in the mold while it is cooling, stresses left in the part as it cools are severe. A look at a transparent or translucent part under polarized light will show this well.

A part with molded-in stress will tend to relieve itself when exposed to high ambient temperatures or incompatible chemicals. Such parts will warp or deform causing many dimensions to go out of tolerance. Stress cracking around stress points (sharp corners on the parts themselves or sharp points or edges on metal inserts) will tend to develop, causing failures in assemblies or fastening areas.

In structural foam, the fact that less pressure is used to fill the mold reduces this stress level dramatically. Parts will tend to stand up to heat and chemical attack much better than an injection-molded part. Inserts tend to hold better, and stress cracking is less likely. Physical properties are also apt to be high due to this lack of stress, and impact strength is generally improved. The energy absorption characteristic of the cellular product may also have something to do with this.

V. APPLICATIONS FOR CONSIDERATION

Although structural foam has found many uses in many areas, the fact remains that there are really only two main use areas. These two groups—enclosures and material handling uses—are quite broad. The enclosure area involves mainly the electronic industry and is satisfied by a whole series of so-called engineering materials, all of which are flame retardant in some manner.

The material handling market uses mainly the commodity resins and, at least at the present time, requires no flame retardant materials. There are, of course, exceptions.

The following photographs show representative samples of structural foam applications:

FIGURE 2.1 One of the hundreds of business machine enclosures developed over the last few years. This small business computer workstation is made in General Electric's NORYL-modified P.P.O.

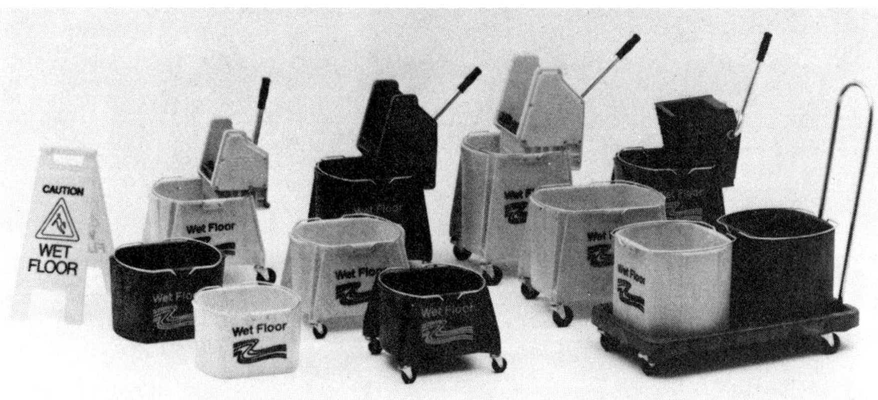

FIGURE 2.2 This popular product probably represents one of the toughest uses of structural foam today. Developed and marketed by Rubbermaid Commercial Products, Inc., this line of utility mop buckets and wringers, is produced in high-density polyethylene.

FIGURE 2.3 This is a photo of one of the new TV satellite antennae disks now being produced in structural foam. This model is molded of high-impact polystyrene by Poly Plastic Products in California.

FIGURE 2.4 One of the older structural foam applications, this heavy duty milk case is still being utilized by dairies around the country. The base material is high-density polyethylene and the units are produced in several colors with the dairy's logo hot-stamped in the side panels.

FIGURE 2.5 Unicell cabinet system.

FIGURE 2.6 Road barrier.

REFERENCES

1. Three Ways to Produce Desktop Computer Housing, *Plastic World*, Sept. 1982, p. 34.
2. Three Ways to Produce Desktop Computer Housing, *Plastic World*, Sept. 1982, p. 36.
3. Three Ways to Produce Desktop Computer Housing, *Plastic World*, Sept. 1982, p. 38.

3

From What Is Structural Foam Produced?

I. INTRODUCTION

There are many materials available from which to produce a structural foam. Two basic forms, thermoplastic and thermoset, offer a choice in production techniques. Each of these has families of polymers with their own set of properties. The choice of which material becomes more difficult as combinations of these families and fillers of all types become available.

II. MATERIAL GROUPS—THERMOPLASTIC VERSUS THERMOSET

When looking at the materials from which to produce structural foam, we generally break all available materials into two groups. Thermoplastics, which can be recycled, and thermosets, which cannot. The thermoplastics normally require no chemical reaction when being molded; the thermosets do in fact react as they are being molded. As you might expect, the thermosets usually have higher physical properties over a broad temperature range.

In this category we include urethanes, epoxies, polyesters, ureafomaldehydes, and many others, which are processes in various ways. The most popular being reaction injection machines or RIM, as it is known. We will cover this process in more detail later.

For the purposes of comparing these materials, we include below a property chart of these products. It is obvious that formulations for these systems are virtually unlimited. Provide a list of properties you require and you can no doubt design a thermoset system to meet them. Problems still exist, however, in processing and economics. Mold release can be troublesome and ambient conditions such as humidity and temperature can play havoc with a system. Also with a chemical reaction on a production line, there is a large potential for problems.

There is very little doubt that an in situ polymerization of a plastic product is the process of the future, but how far away is anybody's guess. It has always been more practical to bring chemicals to the molding machine in drums, react them in an extruder/ reactor, and inject them into the mold. A simple control on the metering system can change the properties from rigid to elastomeric, from solid to foam, or whatever.

In the meantime, the simple processing and inexpensive costs of thermoplastics make them the "now" product. In recent years, new polymers with very exciting properties have been developed. Alloys of two or more polymers and additives of glass fibers, carbon fibers, metal fibers, and others have made it almost impossible to keep up with available products.

A. Thermoplastics

The thermoplastic materials used for structural foam normally end up in two categories: the so-called commodity resins, which include high-density polyethylene, polypropylene, and high-impact polystyrene, and the others, which seem to end up in a misclassification known as "engineering resins." This group includes polycarbonate, polyphenylene oxide (Noryl®), ABS, thermoplastic polyester, and a myriad of others. By combining one with another or by adding additives, we can tailor-make a resin system to do a specific job and so develop an "engineering resin."

Properties relating to these products in the foamed form are not very complete and are often tested under a variety of laboratory conditions. This makes it very difficult to design using standard design equations. One must keep in mind that a foamed product is never stiffer nor stronger than the same cross section in solid form. When you take material from the center of a part, you do in fact reduce properties.

Most resins used for foam will give you a strong enough structure to sustain most use requirements. Such factors as use temperature, UV resistance, and resistance to other environmental conditions

TABLE 3.1 Materials

Generic Name	Cost Range[a]	Typical Application Area
High-density polyethylene	L	Material handling items
Polypropylene	L	Material handling, furniture
High-impact polystyrene	L	Furniture
High-impact polystyrene with flame retardant	M	Electronic enclosures
ABS	M	Industrial parts
ABS with flame retardant	M	Electronic enclosures
Modified polyphenylene oxide	M	Electronic enclosures
Nylon	M-H	Commercial items
Polycarbonate	H	Electronic enclosures—high heat
Rigid urethane	M	Enclosures, commercial items

[a]L = lower; M = moderate; H = higher.

TABLE 3.2 Thermoplastic Structural Foam Properties (Physical Properties Guidelines at 0.250 Wall with 20% Density Reduction)

Property	Unit	Method of Testing	High Density Polyethylene	ABS	Modified Polyphenylene Oxide	Polycarbonate
Specific gravity		ASTM-D-792	0.6	0.86	0.85	0.90
Deflection temperature Under load	°F	ASTM-D-792				
@66 psi			129.6	187	205	280
@264 psi			93.5	172	180	260
Coefficient of thermal expansion	in./in./°F × 10^{-5}	ASTM-D-696	12	4.9	3.8	2.0
Tensile strength	psi	ASTM-D-638	1,310	3,900	3,400	6,100
Tensile modulus	psi	ASTM-D-638		250,000	235,000	300,000
Flexural modulus	psi	ASTM-D-790	120,000	280,000	261,000	357,000
Compressive strength 10% deformation	psi	ASTM-D-695	1,840	4,400	5,200	5,200
Combustibility rating		UL Standard 94[a]		V-0[a]	V-0/5V[a]	V-0/5V[a]

TABLE 3.2 (Continued)

Property	Unit	Method of Testing	Thermoplastic Polyester 30% glass filled	Poly-propylene	High Impact Polystyrene	High Impact Polystyrene with FR
Specific gravity		ASTM-D-792	1.2	0.67	0.70	0.85
Deflection temperature						
Under load	°F	ASTM-D-792				
@66 psi			405	167	189	
@264 psi			340	112	176	
Coefficient of thermal expansion	in./in./°F × 10^{-5}	ASTM-D-696	4.5	5.2	9.0	
Tensile strength	psi	ASTM-D-638	9,910	1,900	1,800	2,300
Tensile modulus	psi	ASTM-D-638	1,028,000	79,000	141,160	245,000
Flexural modulus	psi	ASTM-D-790	1,000,000	80,400	200,321	275,000
Compressive strength 10% deformation	psi	ASTM-D-695	11,300	2,800	3,447	
Combustibility rating		UL Standard 94[a]	V-0[a]	HB[a]	HB[a]	V-0[a]

Material properties given above are typical and vary from supplier to supplier. It is recommended that an end user contact his supplier and/or molder to obtain specific properties for use in a given application.

[a]This rating is not intended to reflect hazards presented by this or any other material under actual fire conditions.

Properties above from Structural Foam, S.P.I. Structural Foam Div., 1984.

will dictate what resin to use. One should always use the least expensive resin that will do the job. All plastics applications are compromises, and it is unlikely you will be the first to find a perfect plastic for a given application.

1. Shrinkage

Remember that each resin has its own mold shrinkage. Some are grouped close enough together so that a mold built for one will produce the same part size in another. Remember also that additives such as glass will change this shrinkage.

Consult your molder when making a selection and pick at least two or three materials that can be used. Start your evaluation with the most obvious and go to the others if you encounter problems.

2. Foamed Grade Resins

Many material suppliers provide specific grades of resin for foam molding. In most cases, these resins are lower in cost than their solid injection-molded counterparts. Some foam grades of resins, such as polycarbonate, contain small percentages (5–6%) glass fiber to act as nucleating agents. These particles help small uniform bubbles to form as the blowing agent comes out of solution.

Since most foam parts have fairly thick wall sections, tight specifications on molecular weight are not needed. Wide-specification material not usable in injection-moldable grades can be used and this allows material suppliers a means of getting rid of otherwise nonsalable product. One can use most grades of molding resin to produce foam, therefore it is important to be specific when requesting a particular resin. Generic names, such as polystyrene, may not be adequate.

3. Regrind Materials

Since thermoplastics can be reused several times before properties are affected, it is good economic sense to allow your molder to use some regrind. Up to 20% can be used in structural foam without seriously affecting properties.

Some applications requiring the best physical properties possible may require only virgin material. This will increase the cost, but will also assure you of getting no regrind. A pressure vessel in polycarbonate might be such an application.

OEM often require that their molders certify the material being used in their parts. This is normally done by requiring molders to receive a log-by-lot certification from the material supplier and then follow a closely controlled material logging system as the material is molded.

Some agencies, such as CSA in Canada, have from time to time requested actual samples of each lot of material to be shipped and kept on record for follow-up testing if required. These procedures often add costs for the molder, but most keep lot-to-lot logs on their daily production anyway.

4. Thermoplastic Structural Foam Properties

Physical properties of various engineering materials in the structural foam form are given in the appendix.

B. Thermosets

The most popular of this type of material is a family of products based on urethane chemistry.

1. Production of Rigid Urethane Foam

Polyurethane structural foam produced by the reaction injection molding (RIM) process makes possible the manufacturing of large-area, thin-walled, load-bearing components more simply, more economically, and faster than by conventional manufacturing methods.

Polyurethane structural foam is a sandwich structure consisting of two solid nonporous skins with a gradual transition to a low-density microcellular core. This versatile sandwichlike material, which is molded in a single shot, offers the widest range of properties in the structural foam industry today. Specific gravities between 0.4 and 0.9 are easily attainable whereas flexural moduli can range from 10,000 to 350,000 psi. RIM structural foam materials can even be combustion modified for a UL-94, V-0 rating.*

Since the RIM components enter the mold as a liquid, parts can be produced that vary in thickness from 1/8 to 2 in. without sink marks. Similarly, a wide variety of composites can be molded by foaming directly onto substrates of almost any material, or by encapsulating various reinforcements.

---

*UL-94 is a testing procedure for electrical appliances and devices and incorporates a small-scale test for combustibility that does not necessarily reflect the performance of the material under actual fire conditions.

Physical and Chemical Properties: A variety of physical and chemical properties exist for polyurethane structural foams, depending on the particular chemistry chosen. For instance, a comparison of two systems in Table 1 illustrates the diversity of physical properties attainable with RIM urethane chemistry.

In addition to the particular formulation chosen, RIM structural foam properties are dependent on density and thickness. Table 2A shows the effect of density on a polyurethane structural foam's properties, whereas Table 2B compares the same system's properties at different thicknesses. As is indicated in Table 2A, RIM structural foams are normally molded in the 0.4 to 0.6 specific gravity range.

Polyurethane structural foams have some unique properties as compared with other structural foams because of their density and sandwich structure and because they are thermosets.

Chemical Resistance: Molded parts manufactured from polyurethane structural foams have excellent resistance to more solvents and other potentially damaging materials such as water, salt water, aliphatic hydrocarbons, and a large variety of commercial solvents.

Weatherability: Polyurethane structural foams are resistant to aging and weathering, although the sun's UV rays may darken the surface in outdoor applications. The end use of the part should determine whether or not the surface needs a protective coating.

Surface Appearance: Polyurethane structural foams, unless specially pigmented, are tan in color. RIM structural foam faith-

TABLE 3.3 Physical Properties Attainable with RIM Urethane Chemistry

		System A	System B
Specific gravity		0.5	0.85
Flexural strength	psi	4,800	6,500
Flexural modulus	psi	100,000	240,000
Tensile strength	psi	1,700	5,000
Heat deflection temp., °F @ 66 psi		196	222
Charpy impact, unnotched, ft lb/in.^{2}		8.6	25.0

TABLE 3.4A Typical Physical Properties

Property vs. Specific Gravity			
Specific gravity	0.4	0.5	0.6
Flexural strength (psi)	3,500	4,600	5,200
Flexural modulus (psi)	86,000	106,000	121,000
Tensile strength (psi)	1,400	1,800	2,400
Heat deflection temp., °F @ 66 psi	178	196	205
Charpy impact, unnotched, ft lb/in.^{2}	6.3	8.6	10.0
Skin hardness, D scale	70	75	80

TABLE 3.4B

Property vs. Thickness	1/4 in.	3/8 in.	1/2 in.
Specific gravity	0.6	0.6	0.6
Flexural strength (psi)	5,400	5,400	5,200
Flexural modulus (psi)	156,000	143,000	121,000
Tensile strength (psi)	2,600	2,500	2,400
Charpy impact, unnotched, ft lb/in.^{2}	12.0	11.0	10.0

fully duplicates the mold surface. There are no swirl marks, thus textured, wood grained, or smooth glass-like surfaces can be reproduced to meet end-use design requirements.

Accoustical Properties: Because of their sandwich structure, polyurethane structural foams can surpass the sound-deadening characteristics of wood within the audible frequencies. Again, this property depends on the density and thickness of the part.

Combustibility: Polyurethane structural foams can be formulated to obtain different degrees of combustion resistance. Parts can be molded to meet the V-0 and 5-V combustibility rating when tested in accordance with UL-94* [2].

2. RIM—A Process Whose Time Has Come

Dramatic advances in material technology are making these exciting times for RIM. Internal mold release technology finally appears to be practical, boosting productivity by as much as 50%. Conventional RIM materials are cycling faster, while light-stable systems promised to control the old color-fading problem. Nylon 6 will become available in improved grades, further boosting the prospects for this RIM material. These developments are the vehicle that will carry RIM processors to an estimated quadrupling of market volume in the next 5 years. Specialty grades of RIM and nonurethane RIM materials are proliferating at almost alarming speed. So far, only nylon has made the breakthrough to full commercialization.

All of these materials can be foamed as they are produced. This new line-up of resins processed on the RIM machines offers the designer many new opportunities in the structural foam arena.

III. REGRIND VERSUS VIRGIN MATERIAL

Once a specific material has been chosen for an application, it should be determined how much regrind will be allowed to be used when molding the part. Structural foam equipment, which is normally extruder fed, is most forgiving and more regrind can be used without affecting physical properties. Where wall sections are 1/4 in. or thicker and the application does not require highest physicals, it may be more economical to allow the molder greater latitude in his choice of mixture.

For engineering materials such as polycarbonate, the problems with moisture and contamination can cause severe loss of properties. However, a 1/4-in. piece of polycarbonate structural foam is quite strong and may well be acceptable in the application. This would not be true of a thin-walled injection-molded part. The rule of thumb in injection molding is usually only 10% regrind. In foam, up to 20% could be acceptable.

---

*Waiver.

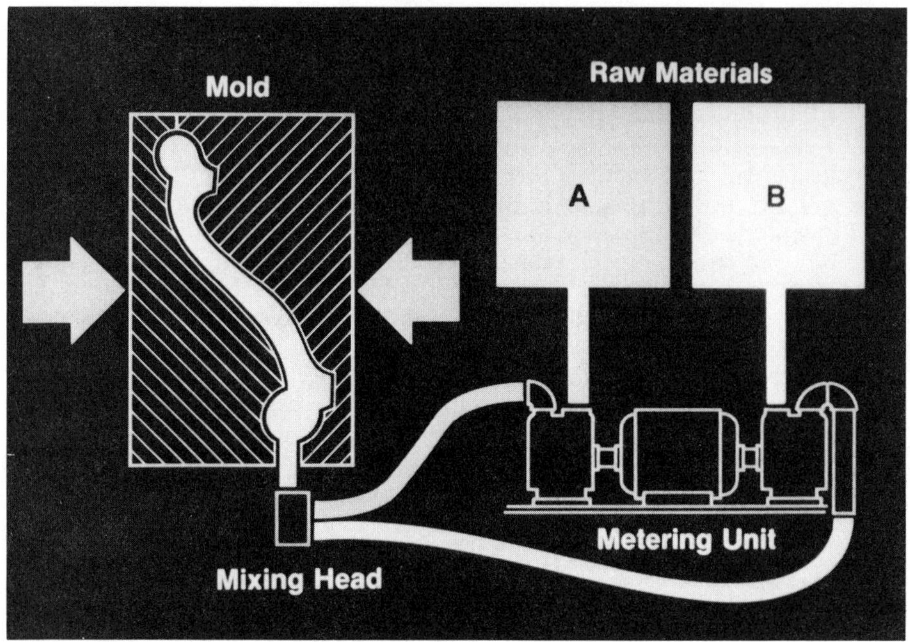

FIGURE 3.1 Diagram of metering and mixing system for RIM processing.

REFERENCES

1. James Eakin, Thermoset Foam Production, *Structural Foam*, Society of Plastics Industires, 1984, p. 21.
2. Agostino von Hassell, RIM Materials Take Quantum Leap Forward, *Plastics Technology*, March 1983, p. 37.

4

How Is Structural Foam Produced?

I. INTRODUCTION

A review of the processes involved in producing foam from both thermoplastic and thermoset is needed. Each has its own problems, techniques, and advantages. All are in a state of change and must be monitored constantly to keep track of improvements both in properties and economics.

II. PRODUCTION OF THERMOPLASTIC STRUCTURAL FOAM

In producing thermoplastic structural foam, the polymer is heated to liquid form and mixed with a gas, such as nitrogen, or a chemical blowing agent which under heat breaks down into ammonia or other gas. Material is mixed together in a liquid-gas or liquid-liquid system under pressures of 2000–10,000 psi. The material is then moved through a system of manifolds, through the nozzles, and to a chilled mold cavity, into which it is injected at relatively high speeds.

The material then expands due to decrease in pressure and immediately starts to foam and fill the cavity. The first layer of plastic and gas bubbles hit and break against the wall of the mold, forming a skin. This system soon reaches equilibrium and a structural integral skin with a cellular core is produced.

Because of the insulative characteristics of this material, it tends to hold heat and the part must be kept in the mold until the skins build sufficient strength to withstand the continuing expansion tendency of the plastic-gas mixture in the middle of the part. Should the part be taken out of the mold too soon, the hotter center of the part will soften the already chilled skin allowing the residual gas pressure to push the heat-softened walls outward forming a bulge. This is exactly the opposite of injection molding in which the material tends to shrink as it cools, leaving recessed areas immediately opposite ribs and thick sections.

A. Improvements

Over the last 15 years, various techniques and equipment to produce different types of structural foam have been developed. Improvements demanded by the customer include:

1. High-gloss smooth surfaces
2. Lower weight
3. Reduction in wall thicknesses to below .100 in. with economic tooling
4. Reduction in overall price of the material in the molded form

It is very difficult to achieve any one or a combination of these without seriously affecting one of the others.

B. Equipment

In the processing of thermoplastic materials into structural foam, many different types of equipment and processes have been developed to provide different forms. The early foams were produced in styrene, using chemical blowing agents. Later, nitrogen was mixed with the thermoplastic in the extruder barrel. As the process caught on, many different types of equipment were developed to provide different means of foaming and forming.

1. Low-Pressure—Multinozzle

The lower-pressure, multinozzle system developed early by Union Carbide is still probably the single largest technique for producing thermoplastic structural foam. This equipment uses large vertical or horizontal hydraulic presses, with an extruder-fed manifold system. Normally, the nitrogen gas is injected into the extruder barrel half way down the unit between a two-stage screw.

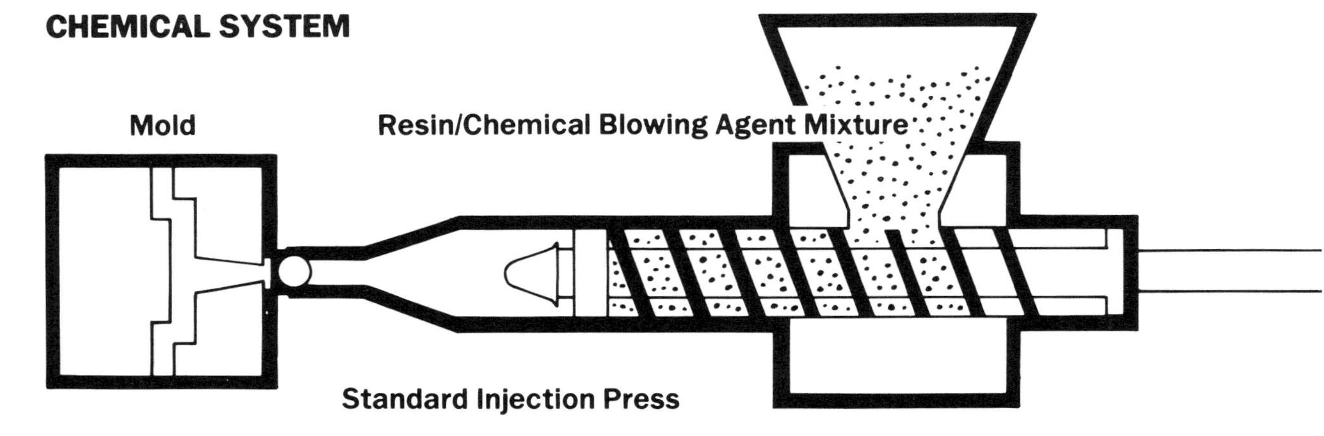

FIGURE 4.1 Diagram of standard injection molding machine utilizing chemical blowing agent to produce structural foam parts.

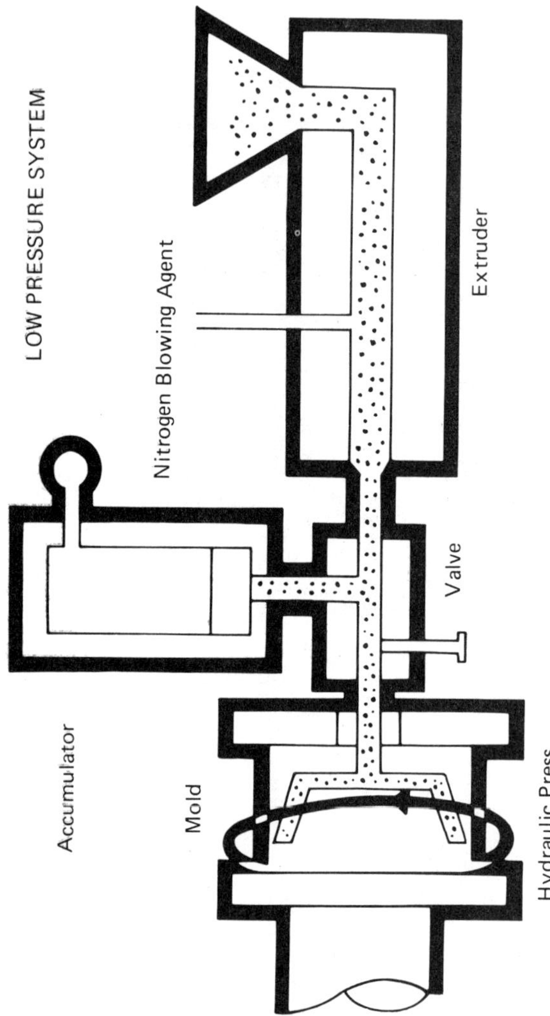

FIGURE 4.2 Diagram of low-pressure system utilized on most multinozzle foam machines.

FIGURE 4.3 One of the larger structural foam machines ever built, this 500-Ton Hoover Universal multinozzle has a 100-pound shot capacity. The Ford automobile parked under the clamp portion gives an idea of the size of this piece of equipment.

At this point, several flights of screw are eliminated to develop a pressure drop. The second stage of the screw mixes the nitrogen and thermoplastic into a homogeneous melt. This material mix is then fed into an accumulator.

Once a predetermined shot size is captured, it is injected through a manifold system into the mold through syncronized nozzles. The machines usually range in tonnage from 150 to 500 and have shot capacity from 20 to 150 lb. An example is shown in Fig. 4.3.

The biggest single advantage of the multinozzle low-pressure system is that nozzles can be arranged into any area of the part and large parts can be produced without long runs of material. High pressure with the consequent high stress level is thus reduced.

2. Low Pressure—Single Nozzle

Low-pressure, single-nozzle equipment has also been used over the years. This is due to the availability of injection-molding machines designed specifically for solid melt which have been utilized to produce structural foam. These machines utilize a screw and barrel system for both plasticizing and as an accummulator. Normally, a chemical blowing agent is utilized with this type of equipment.

The blowing agent breaks down into a gas such as carbon dioxide or ammonia and is in turn mixed in the melt through normal plasticizing. The material is then injected into the mold through a single nozzle.

The short-shot technique is utilized, which means the material is pushed into the mold until it is only about 80% full. The expansion of the gas then fills the remaining portion of the mold causing the skin effect and the foamed core structure. The main disadvantage of using the low-pressure, single-nozzle system is that the equipment does not have the capacity for large parts and the press usually has much higher clamp pressure than required.

3. High Pressure—Single Nozzle

High-pressure, single-nozzle systems designed exclusively for structural foam have been developed, mainly in Europe. These systems are finding wide use in the United States. Higher injection pressures are used on low-pressure presses designed for foam. These higher pressures allow the material to be injected at a much higher rate. There are many advocates of this technique who feel that the cell structure is much better and that the part itself has better surface definition. Thinner sections are also possible. An example is shown in Figure 4.4.

4. USM Process

Some off-shoots of this approach have been the USM technique where material is injected into a steel cavity at extremely high pressure. Skins are formed as the material is forced into the cavity from the machine. Once the material skin has formed, the mold cavity is allowed to expand hydraulically, thus reducing the pressure and allowing the blowing agent to expand accordingly.

This provides a structural foam part having very good skin definition and extremely good foam quality. The disadvantage is the higher-priced tooling needed for the higher pressures and the expansion capability.

Although there are other techniques and approaches to producing thermoplastic foams, most are variations of the three techniques mentioned.

5. Structural Web Process

Many new systems under consideration are variations of the above. One is known as the structural web process. This process is so named because of the part's interior configuration. The idea behind the process is to inject gas into a molten polymer in the mold such that the gas-polymer interface is deformed into a wave-like corrugation, using the principle of the hydrodynamic instability of viscous fingering.

The structural web is characterized by an integral skin and a generally hollow center containing a multitude of skin-connecting webs [1].

6. Coinjection

Another new process seeing some use in the United States is coinjection. Sometimes known as sandwich molding, the process involves filling a mold from two plasticizing units each with a different material. The solid material will be put into the mold from one unit, a split second before the foamed plastic is injected from the other.

The resultant molded product has solid-injection-molded skins with a foamed cellular core.

Over the years, several attempts have been made to control the pressure in the mold cavity. By keeping cavity pressures higher than ambient, the blowing agent does not come out of solution until the melt mix is in the mold and skins have started to form. This technique produces a good smooth skin and a uniform cell structure.

A gasketing of the mold cavity and a back pressure of inert gas is required. This usually adds cost both to the mold and the equipment.

Allied Chemical developed a process similar to this in the 1960s and others are working on improvements today.

C. Thin Wall Versus Standard $\frac{1}{4}$ in. Wall

The discussion continues over the value of going to thin wall foam. On the basis of present methods of molding, this falls into the range of 0.150–0.180 in. Pushed by a few material suppliers, this supposedly new technique is really nothing new.

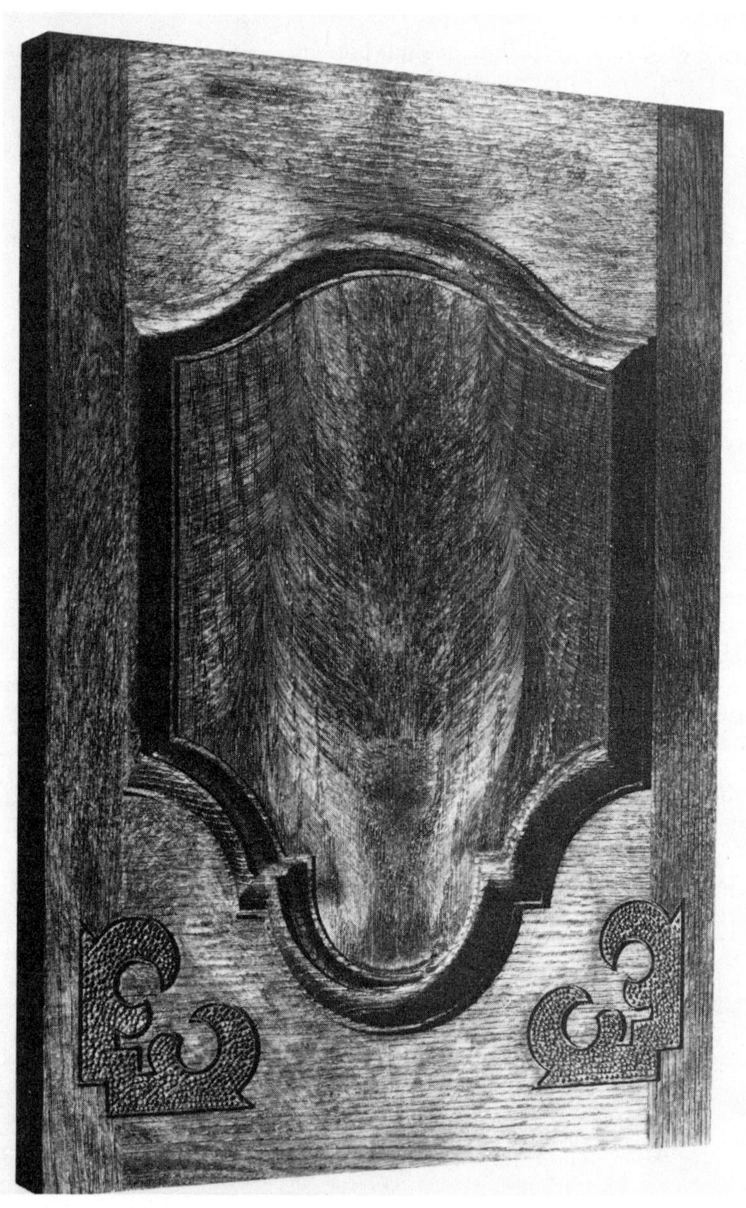

FIGURE 4.4 With the higher pressure foams one can obtain excellent woodgrain detail. This panel molded of high impact styrene is used as a door for an expensive piece of furniture.

By offering some lower-molecular-weight materials and increasing the molding injection pressure, the thinner walls are much easier to obtain. The increased pressure does, in fact, improve surface condition, thus reducing the preparation time for a part to be made ready for finishing.

The danger of damaging tools also increases and it would be well to consider steel molds when taking this approach. One should also give serious consideration to going all the way to solid injection molding. With a small amount of blowing agent added, shrink can be reduced and painting may not be necessary at all. For over 30 years, molders have been adding small amounts of blowing agent to reduce sink.

At the lower cost, a result of faster cycles, and no paint, the solid or near solid part is a much better bargain. One must be sure structural foam is really needed based on application requirements, not just because a new process or technique is available.

D. Chemical Blowing Agents

The production of a structural foam part requires a source of gas to provide a cellular structure. Much of the low-pressure structural foam equipment is designed to use compressed nitrogen gas as the foaming agent. Chemical blowing agents (CBAs) may also be used in the equipment, however; there are also several low- and high-pressure structural foam machines designed to use only CBAs as the gas source. Where conventional or modified conventional injection molding machines are used to produce foamed products, a CBA is required.

Several color specialty companies have developed alternative forms of CBAs to improve their handling characteristics. These include liquid dispersions that are pumped directly into the feed throat of molding machines. Pelletized CBA concentrates are available in a number of different base resins, and flake and bar stock forms of CBAs reduce or eliminate dust and allow for better utilization of automatic metering and blending equipment.

III. PROCESSING THERMOSET STRUCTURAL FOAM—RIM [2]

Polyurethane structural foam is produced by reaction injection molding (RIM). This process involves the metering and mixing of two reactive components, an isocyanate and a polyol. The two chemicals are pumped under high pressures, at a specific ratio, through a self-cleaning mixhead. The mixhead, which is fixed to the mold, statically combines the two reactive components by high-pressure impingement. The low-viscosity reactive mixture is then injected into the mold at atmospheric pressure to between 30 and 85% of the cavity volume, depending on the part density desired.

As the liquid mixture enters the mold and begins to react, the exothermic heat generated vaporizes a blowing agent, thereby causing the mixture to expand and fill the mold. Pressures between 30 and 90 psi are generated within the mold as the cavity is filled. This pressure, in combination with the cooling effect of the mold wall, causes a high-density skin to form on the mold surface, which gives the material its sandwich-like structure.

The surface quality of a RIM part is dependent on the quality of the mold. Specifically, the mold surface is reproduced on the final molding. In addition, temperature control of the mold is important and production should, therefore, only be run in temperature-controlled metal molds. Aluminum has proven to be an excellent mold material, but steel is recommended for high-volume production.

Since RIM involves low molding pressures, low-tonnage clamping units or self-contained molds are utilized. Often these units must be positioned at an angle so that during foam expansion air can escape from the mold cavity.

RIM mold cycle times are influenced by several factors, including the type of structural foam system utilized. The part configuration, wall thickness, and density all are important factors. Typically for a part $\frac{1}{4}-\frac{1}{2}$ in. thick, at a 0.6 specific gravity, a 2-min cycle time would be expected.

Processing equipment for RIM structural foam has the advantage of lower capital investment as compared with thermoplastic structural foam. Lower-cost tools can be utilized with RIM since the low pressures generated in the mold do not require as high a clamping pressure. In addition, one RIM dispensing unit can automatically service six to eight molds.

As previously stated, the RIM reactants enter the mold as a liquid. Design latitudes are wide, therefore, because complex shapes, surface texturing, variable wall thicknesses, and ribs are easy to fill with the liquid mixture.

REFERENCES

1. Olagoke Olabisi, Structural-Web Molding, *Plastics Engineering*, Society of Plastics Engineers, Inc., October 1983, p. 25.
2. *Structural Foam*, S.P.I. Structural Foam Disivion, 1984.

5

How to Purchase Structural Foam

I. INTRODUCTION

Once the decision to consider structural foam as a material of construction has been made, the need to choose a vendor is critical. Unless you have a favorite molder with whom you have worked closely before, it may be necessary to pick several from which to solicit quotations and suggestions. A wealth of information is available from these molders and much of it can save your company dollars.

II. CUSTOM MOLDING

As a purchaser of structural foam parts, you will probably be dealing with what is known in the trade as a custom molder. These companies usually provide to their customers the service of molding plastic parts on customer-owned tooling. Along with this molding service comes a lot of other services such as secondary operations, finishing, and some assembly. One also relies on his molder to provide some consultation on plastic material selection and design. It is normal for the molder to also be the customer's agent for the purchase of a tool. Most molders do not have their own tooling facilities, but they do take responsibility for the design and construction of the tool.

Some molding shops also have proprietary products which they produce and sell. There is often concern on the part of customers buying from a shop that does both custom and proprietary molding. They feel that somehow they will be neglected if there is a question between which parts get molded. Since this can also happen between two custom jobs, it appears that one should be more concerned about the molders' reputation for keeping his word.

In the up and down business of custom molding, it is helpful for a molder to have some products of his own to run when custom business is slow.

III. THE DESIGN-PURCHASING RELATIONSHIP

The design of a structural foam product is much like the design of any plastic item. A blend of functionality and appearance is necessary for any product. Thus, the task assigned to a design group developing a new program is not an easy one. Politics usually plays an important role.

According to an article in *Plastics Design Forum*, [1] two designers assesed it this way:

> "Marketing usually will want a product to have maximum possible salable product features, manufacturing wants everything to be simple, quality assurance personnel . . . want guarantees, purchasing wants the lowest cost. Management wants all of the above on a shortened delivery schedule. It is difficult, if not impossible to placate all of these diverse interests."

We have all played a part in the above scenerio from time to time. Structural foam projects are no different than the one described above.

A. The Design Team

Again according to the *Plastics Design Forum* article, [1] "Much of today's product development work is done under a management team dominated by "bottom liners" who demand a quick return on investment. Great pressure is placed on design engineers to create the lowest-cost, easiest-to-produce new product in the shortest possible time." The choice of structural foam can make a significant contribution toward this goal if the design team recognizes the advantages as well as the disadvantages.

"Design details that are tailored to suit a specific material and process can mean the difference between success and failure," according to the *Plastic Design Forum* articles, and most structural foam molders would agree.

The design team can save many hours of work for purchasing by bringing the molder into the design picture early in the project. Relative costs for tooling, piece price, and time estimates from an experienced molder could spell the success of the project. This information is usually free for the asking and it seems ridiculous, if not outright incompetent, not to take advantage of it.

B. The Buyers Role

In evaluating this from a buyer's point of view, Milt Montz, a buyer himself with Beckman Instrument, a California medical electronic house, has the following comment.

> "One of the buyer's roles is to bring the designer together with sales engineers of the various processes to ferret out which process should be considered for the particular application in question. This action must also take place in the early stages of the design due to the differences in final part design from process to process. The buyer should use preliminary drawings to request formal bids for all feasible processes so he can do a cost/benefit analysis on each. Together, the buyer, the designer, and the manufacturing engineer must agree on the right process, in this case, structural foam" [2].

IV. MOLDER SELECTION CRITERIA [3]

It is important to know where to start looking for potential molding sources. Obviously, a list of molders is necessary, but one cannot just randomly go out and put together a list. Lists of possible molders can come from several different sources.

Look at your competitors and find out who is doing their molding.

Material suppliers are always good sources of molders since they sell to them direct.

Sales representatives selling any type of plastic and who are now calling on you can possibly give you a list of people who they think may be good potential sources of foam molding. Plastic molders in the field who are now supplying your injected molded parts, conceivably know who is doing what in structural foam. There are also published lists of foam molders.

A. First Response

Once a list has been made and you have established initial contact with each of these molders, first responses are always very important. A sales representative, who gives you good follow-up and service and has a grasp of the markets into which he is selling, cannot only lead you to a good molder, but probably can supply you a lot of back-up information as well. Although it is not necessarily true that good sales representatives always represent good molders, you will generally find most sales representatives are fairly honest about the capability of the companies that they represent.

B. Literature

The second area of consideration should be the company's literature. Going over the literature is usually a good way to get an indication of what that molder is trying to sell.

C. Preliminary Quotations

The third area of early contact that should be examined closely is the preliminary quotation. In most cases, end users go out to a large number of molders to get an idea of what the budgetary cost of a project might be. Some end users tend to want the final price right down to the last penny. Because design information is not always available, it is usually impossible to provide an accurate quote at this time. However, the preliminary quote does give you an opportunity to see how the molder handles the input given to him. Appearance and completeness of the quotation should be considered, plus the turnaround time, always an indication of how the molder is handling this particular commercial relationship. Quotes are the life's blood of most custom molders, whether they mold straight injection or structural foam. Their response to your needs, both verbally and in writing, are an extremely good indication of how well they will handle the business at a later date.

Once you have narrowed the field by picking several suppliers, it is important to review each in more detail.

D. Molder Reputation

The reputation of the molder can be easily determined by checking with material suppliers, existing customers, the molder's competition, and sales reps covering similar lines. This is always a good opportunity to get an insight into how other people see your potential supplier. Don't be afraid to ask these people how they feel about this or that molder. Most will be only too eager to tell you.

According to a paper prepared by Milt Montz, a buyer for Beckman Instrument in California, "The reputation of the molder is of paramount importance. . . . One should not align one's self with a molder until one is satisfied with the molder's reputation in the industry" [1].

E. Evaluating the Quote

The evaluation of a quote as it finally comes in is a second indication. Look at not only price completeness, but consider questions asked concerning information not supplied, or misunderstood, completeness of quote, and whether all questions are answered. Obviously, appearance is an extremely important factor. Compare exceptions taken. Most good molders will spot a problem area.

According to Milt Montz of Beckman Instruments, when asked how molders felt about his review of their quotes he said, ". . . it was mentioned that most molders believe that many buyers know little about the structural foam process. Therefore, since buyer knowledge is limited on the process, many molders interviewed expressed concern that their bids on many programs are not adequately evaluated" [2].

F. Vendor Calls

The technical background provided during a visit to your company by a potential vendor prior to you making any type of a plant survey is also important. If sales people who come into your facility are knowledgeable and ask the right questions, you can be reasonably sure that the vendor's follow-up and ultimate production capabilities will be equally as good.

G. Plant Survey

Once you have narrowed the field to a reasonable number of potential suppliers, it is time to pack your bag and head out for a plant inspection. A check-out sheet and a scoring system that can be used in taking a look at these various custom molders is included. To some, "gut feeling" is all that's required, but others tend to want to be more organized. Score molders on various aspects of the visit. No matter how you approach this problem, the same questions should be asked, and the important areas looked at before making the final selection.

Milt Montz, at Beckman Instrument, utilizes seven key people to make a survey. "The vendor survey allows a team approach where an engineer, buyer, manufacturing engineer, and a quality assurance representative all visit potential suppliers and evaluate many aspects of their business," he notes [2].

1. Plant Appearance

When looking at the plant itself, the general appearance will tell you a great deal about how well the company is doing. Does the overall appearance indicate a type of company you are interested in dealing with?

2. Material Flow

Inside, one of the most important areas to check is the lay-out of the equipment and material flow. Ask your host to give you a quick run down of how the material flows through the plant. Ask for a plan showing the way material moves from one end of the plant to the other and finally out as finished goods. Ask him to sketch it out for you if it is not readily available.

3. Equipment

Check the equipment over. Is it old? Well kept? Is there a standard preventive maintenance program? Don't be afraid to talk to the maintenance people concerning their approach to keeping the equipment running. Most structural foam machines are designed to run 24 hr/day, but it is extremely important that the maintenance on these very expensive monsters be kept up. A down machine provides no revenue for its owner, and more importantly, no product for you.

4. Cleanliness

Cleanliness is certainly an important factor and one that should be rated critically when visiting a structural foam molder. Although it is not necessary that you be able to eat off the floor, it is necessary to have good housekeeping practices. Structural foam machines have a tendency to give off a lot of hydraulic oil, cooling water, and general plastic scrap. These things constantly have to be cleaned up. It is not unusual to find even the cleanest of structural foam plants getting pretty dirty over a 24-hr operation. It is therefore necessary to stop and clean up on a fairly regular basis. Obviously, there are going to be times when areas are not as clean as they should be. This should be taken into consideration.

However, one can get a pretty good idea of how well the plant is run by just looking at how well it was cleaned up for your visit. If it is messy when you go through on a tour, even after ample advance notice was given of your visit, you should ask why?

5. Auxiliary Departments

Auxiliary departments are also key when it comes to producing structural foam. Because of the nature of structural foam, a lot of work such as the insertion of brass inserts is done in the secondary department along with sanding and preparation for finishing. These areas should be well managed and kept as clean as possible with a good material flow pattern.

Finishing: The finishing department should also be looked at from a material flow standpoint. Parts going down the line should be checked since their appearance is a good way to find out whether the molder you are looking at does, in fact, supply a quality painted part.

Quality Control: Quality control and the importance a molder puts on it in his plant is probably the number one area of concern with most end users. Unfortunately, with a lot of electronic houses, quality control is a much different animal in their plants than what is required in a standard structural foam-molding operation. When most electronic manufacturers think of quality control, they think of elaborate laboratories, white coats, technicians, and a massive checking, inspection, and metering operation. Unlike electronic industries, structural foam molders tend to do most of their quality control on the floor at the machine, in the secondary area, and coming from the finishing line. Parts must look good as well as be within specifications and it is difficult to control this unless inspectors are on the floor at all times watching the product move through the plant. The first line of quality control is the machine operator. All machine operators should know exactly what to look for in the parts that are coming from the machine. Additional inspections should be made regularly at these machines to make sure that nothing has changed in the mold such as a pin being broken, a gouge, or some other malfunction.

Inspection Equipment: As far as inspection equipment is concerned, most structural foam molders have the capability of doing first article inspection, i.e., checking out a part as it comes from the mold the first time. However, other than that, elaborate checking equipment, such as XY coordinate equipment, is generally not available at most small custom molders. This is also true in the area of metallic shielding where most molders provide some type of conductive surface on their foam parts, either as an electrode spray or in solvent-based conductive coatings. Most are not prepared to check conductivity or other highly technical electronic criteria.

Packaging: Checking the packaging of structural foam parts in a molders' plant is also a good way to learn how this potential supplier is going to perform. Talk to the people doing the packaging in the plant and see how they go about wrapping parts and positioning them in boxes. Check the general overall stacking as well as the marking of each box. Remember, the right markings on a box can help assure you of getting the right parts in the right quantities. Errors in part number, quantity, etc., can cause untold problems not only in inventory control, but in your accounting department as well.

6. Tooling Capabilities

Tooling capability is probably the most key area of concern when looking for potential suppliers. Most structural foam molders have sources outside of their own facility doing the tooling design and building custom tools. This list of toolmakers should be discussed with the molder and when possible, an inspection trip made to as many as possible. Find out who does the design work on the tools and who is included in the approval list for these designs.

Tooling Report: Also a report system should be in effect to let you know when you can expect to receive your tool and the various stages through which the mold is progressing. Following the placement of an order and during the time the toolmaker is designing and building the tool, there is a long, quiet, dead space that is often difficult for the end user to comprehend. It is important that the molder keep you abreast of every development in the tooling cycle and let you know exactly where you stand with regards to your tool.

It is unusual for a molder to have in-house toolmaking capabilities, so do not expect an elaborate tool shop in most structural foam operations. Usually, a small tool shop is available for repair of pins, etc., but it is not normally set up to do full-scale tool production.

Engineering Changes: You should also discuss how engineering change orders are handled, either prior to first articles or after the parts have gone into production. A standard procedure for this is necessary if you are going to have good communication between you and your vendor.

7. Personnel

Personnel is another important area that must be considered when reviewing a potential structural foam molder. A custom molder provides you a service and that service comes from his personnel. The people who should be met and talked to during a plant visit include the engineering department, the foreman on the floor, certainly the plant manager and, ah yes . . . even the sales manager. Remember, the sales department in a structural foam operation can provide a great many ideas not only for the part itself, but also the mold.

Technical knowledge and experience learned in the field are both necessary to provide you with the very best information available to make your potential application a successful project. We often say we know more things not to do than anybody else in the business and that is important. It seems that every structural foam application does, in fact, present new problems. Therefore, it is important that you do talk to people who have experience and make sure that you are getting the very best technology available to date.

8. Review of Survey

In review, it is important that four main areas be looked at when evaluating a structural foam molder's plant: (1) material handling within the plant, (2) machine type and maintenance, (3) secondary handling of parts, including painting and packaging, and (4) caliber of personnel.

H. Financial Review

Other items that should be checked when doing a survey include looking at the financial side of the picture. What are yearly sales and what kind of growth has been experienced by the company? What is the capability of the company to add more capacity if required?

Certainly, sources of capital are important, for it is unfortunate, but true, that a good many of our former colleagues who were in the structural foam molding business, have fallen by the wayside simply because they did not have good sources of capital. They were, in fact, unable to keep up with the tremendous investment needed to continue to produce structural foam.

I. Proprietary Activity

It is also important to check other customers that are presently doing business with your potential vendor. Obviously, if companies such as your own are happy with parts received, you can be reasonably sure of the ability of the vendor to produce. Ask your potential vendor for a customer list and his contacts there.

One should also consider proprietary activity within a custom molder. Determine how much of his business is directed toward custom work and how much time proprietary products are going to take up on his machine. Most people feel that a blend of both is important in keeping any plastic molder healthy and aware of what is going on.

J. Long-Range Goals

It is also important to discuss long-range plans with a key officer in the company. Do projected growth pictures look good? Is the company making plans to take care of this growth with new production equipment and space as production requirements increase? A growth company is always better than a stagnant one.

K. Insurance

Finally, it is important to check on insurance that is available. Does the vendor protect your tools while they are in his shop? Are you protected against loss due to a work stoppage such as fire, strike, or a similar situation?

L. The Right Choice

After you have all of this down, it comes right down to deciding what vendor is going to be able to supply you acceptable parts at a cost you are capable of paying. Remember, it still holds that you get what you pay for . . . even in structural foam.

Once you have determined the quality level for which you are prepared to pay, it is equally important to find the right foam molder that will be able to satisfy that requirement.

V. LINE OF RESPONSIBILITY

In the purchase of any product where more than one supplier is involved, a single line of responsibility is a desirable situation. If the product is not right, you want to be able to hold only one party responsible.

In purchasing a structural foam part or set of parts, this is especially critical. In the development of an electronic enclosure, for instance, you may have as many as three suppliers involved in the finished part. The first to be involved is the toolmaker, then the molder, finally the finisher. All play a key part in getting your enclosure "to print" and delivered on time.

If you decide to contract separately, many problems can appear, some with amazing speed; others are more subtle, may take months to develop, and only appear after the housing is securely around an instrument operating in your customer's facility.

A. Tooling

First let us look at the tooling. Most molders will normally contract for the tooling and charge you in the area of 10% to handle the paper work, assist in the design, and, most importantly, take responsibility for the tool. This fee usually includes sampling of the tool to provide first articles. A new mold, depending on complexity, may go into the machine three or four times before the tool is deemed correct and ready for production.

In structural foam, each sampling could cost in the neighborhood of $1200/tool, again depending on complexity. Color matches and other sampling is also done at this time. The bill for testing a mold can be pretty substantial. Purchasing the mold through the molder eliminates all of this and you only pay the tool charge. Usually the last half of that amount is due only when you have received approvable first articles.

B. Finishing

Finishing can include conductive coatings and electroless plating as well. You could purchase the molded parts without finish, bring them into your facility, quality control them, and ship them out to your finisher. Once they're ready, you can quality control them again before putting them into inventory. Sounds simple enough, right?

Here is what could happen. You have brought the parts in, your quality control has approved them, and they go to the finisher. He puts on a conductive coating at approximately $2/ft^2 and then a urethane outer coat at about the same price. You take them into inventory and in 24 hr blisters appear on the surface. You now have a big problem! The finisher will tell you the parts are out-gassing due to improper molding, the molder will tell you the finisher used the wrong solvent. Whose problem? Yours, obviously. A "one-line" responsibility eliminates this problem.

This approach will not eliminate problems entirely, but it certainly makes it easier to address them.

VI. PURCHASING OF TOOLING

The act of buying tooling for a structural foam program is as important as it is complex. The "up front" money usually needed for a project is usually the hardest to come by. There is a long period of time between the down payment (usually 50%) and the delivery of the first article.

There also seems to be a great deal of confusion over what is being purchased, for how much, and why it costs so much.

A. Quality Counts

In the first place, let us put one thing to rest. You get what you pay for. There is a wide range of tooling quality and the price varies accordingly. The biggest problem lies in the lack of knowledge most end users have with tooling in general. They often include it on the spread sheet along with piece price with few if any specifications and wonder why the spread in price.

B. Tooling Specifications

When purchasing tools of any type, it is critical that tooling specifications are included in the "request for quote." If your engineering or design people do not know enough about tooling to specify, you should talk to a toolmaker or a qualified molder and determine the design, type of tool, and material of construction for what you wish to accomplish.

There is no such thing as "cheap" tooling. You get exactly what you pay for. To paraphrase a popular TV ad, "you can pay for it early in quality tooling or later when the parts become more expensive due to mold wear."

A typical specification for structural foam tooling is included below. You can be more specific or you can leave the design in the hands of your molder. However, to obtain a comparative bid, you must define what you want to purchase.

1. A Typical Specification For Structural Foam Tooling

The following specifications are to be followed on all structural foam tooling built for our customer. Any deviation must be approved in writing.

1. Materials of Construction. All material from which cavities or cores are produced must be machined of jig plate aluminum. 6061 or 7075 grades with a T-6 hardness must be used. Slides and ware plates must be either steel or brass. Ejection plates are to be hot-rolled steel, blanch ground, and a minimum of 7/8 in. thick.
2. Actions. May be either cam or cylinder operated, but design must be approved before using.
3. Shunt Height. Shunt height should be between minimum of 12.680 and maximum of 30.680. Height must end in 0.0680; unless otherwise specified, mold should be built to 19.680 typical height.
4. Platen Keys. All molds must have both horizontal and vertical keys on front or stationary side. Drawing for location of keys is attached.
5. Nozzle Location. Nozzles are to be located on 6-in. centers and must conform to layout per platen drawing attached. Location of nozzles should be approved before proceeding with tool design.
6. Nozzle Bushings. All molds are to have steel nozzle bushings and are to conform to drawing attached.
7. Gate Size. All gates should be minimum of 0.200 × 0.500 in. Gates running into edge of part should have sharp, raised, break-off lip on appearance side to facilitate breaking. See attached drawing for detail.
8. Runner Design. Runners should be minimum, 3/8 to 1/2 in., full round with minimum restrictions. Lengths should be kept to minimum.
9. Sprue Design. Sprues should have $2\frac{1}{2}°$ draft toward gate. Surface should be polished. A sprue puller should be used opposite sprue when design demands.
10. Clamp Slots. Slots should be cut on all four sides of both front and back of tool. Surface area, of at least 1 × 3 in. for every 6 in. of mold dimension should be available for clamp retention. The flot should run the full length of the tool, if possible.

11. Eyebolts. Tool should have tapped eyebolt holes on all four sides of tool and on both halves of tool. Diameter depends on mold size: up to 1500 lb, 3/4 in.; over 1500 lb., 1 in. An effort should be made to locate holes in middle of both halves of mold so as to assure its hanging straight when on crane. If design of tool prohibits hole to be in center of mold, two holes on center line should be drilled.
12. Waterlines. Waterlines should be minimum 9/16 in. diameter and tapped for 1/2-in. connectors. Connectors should be included with mold. Distance between lines should be no greater than 3 in. and distance from cavity to center line of waterline should be 1 in.
13. Chain Pulls. When need arises to put cavity side of tool on moving side, a chain pull is preferred over other means. Minimum mold opening is 24 in.
14. Mold Marking and Nomenclature. All molds are to be stamped or permanently marked with owner's name, part name, part number, tool number, if given, and toolmaker's name. Date of completion is also to be stamped on tool.
15. Confidentiality. All drawings supplied to toolmaker either for quotation or mold construction are the exclusive property of the customer. These prints and drawings are not to be shown to any outside parties and all are to be returned after need to have is terminated.
16. Design. All tooling design done on a given mold is the property of the molder and his customers.
17. Tooling Drawings. Two complete sets of tool drawings required with each tool built and will be supplied to customer as well as kept on file at molder.
18. Tool Design. Must be submitted and approved before proceeding with construction.

C. Mold Inserts

A technique often used to build flexibility into structural foam tools is the use of inserts. By this method, several distinct parts can be produced from one tool.

When using this technique there are several things to be aware of. There will be witness lines where two metal edges come together. These can often be removed by sanding and painting, however, this increases price. Also, the normal storage of inserts for a tool often leaves much to be desired. Inserts, especially aluminum ones, often get banged and dented, which causes increased problems with both fitting and witness lines.

If you build structural foam molds with inserts, be sure your mold maker or molder provide wooden storage boxes for each insert when they are not in use.

Oftentimes the problems caused by inserts are just not worth the money you may save. However, properly designed and handled, an inserted mold can give you great flexibility.

D. Identifying Structural Foam Parts

From time to time it is important to determine the source of molded parts either in your inventory or on a finished product. Several techniques are used for this, including a simple date stamp with molder's logo. However, this can be a problem if the parts get painted over after the stamp has been applied.

Another technique is to use a marker in the mold. By changing position with a screw driver, one can have a continuing monitor of the parts being molded from a given cavity and at the same time identify molder, molding date, and lot number. The part number should also be embossed in the mold surface. A revision number can either be embossed at the end of the part number or put in with an ink stamp. All of these features should be discussed prior to placing the purchase order since they do involve some costs.

VII. PREPARATION OF STRUCTURAL FOAM QUOTATIONS

In purchasing foam or any molded part for that matter, it is often helpful to know just how jobs are quoted. Each molder usually has his own formulas for getting to a piece price, but they are made up of the same items.

A. Material Costs

In the first place, material cost is considered. On new programs, since no part exists, the cubic inches of a part must be calculated from the engineering prints. To assist your molder to quote and to get a more accurate comparison between two competitive molders, it is a good idea to calculate the cubic inches yourself and spell this out on the print. You can convert this cubic inch figure to pounds of material if your like; however, two types of processing will give you two different weights. By supplying just cubic inches, you will have all molders quoting on the same basic shape. Factors for converting various materials from cubic inches to pounds are usually available from your molder.

Once the weight of a part has been determined, it can then be multiplied times the cost per pound of material. If you specify exactly what material you want, you should have little difference between quotes. However, if you leave such things as percent regrind allowable unspecified or you call out only a generic name for material, you can find a wide range in material cost alone.

To this material price, molders will add a scrap factor. Ten percent loss is not unusual.

B. Machine Costs

The next step, and often the most misunderstood, is calculating the machine molding cost on the item. Here the design, especially the thickest section of a part, will have a strong influence on the cycle time.

1. Cycle Time

A table for average cycle times is included below, but one must remember that core pulls, fancy shut-offs, and other features will add to this cycle. The part must be kept in the mold sufficiently long enough for cooling in the mold to solidify the skins and make them strong enough to sustain the internal pressure caused by the blowing agent. This is exactly the opposite of injection molding where a sink takes place due to the shrinking material as it cools.

Once a cycle time is estimated, it is divided into a machine hourly rate. On multiple cavity molds, this price per shot is then divided by the number of parts produced per shot.

2. Machine Rates

One is often asked, "What is your machine rate?", as if the answer would provide a means of comparing one molders' prices against anothers. This machine rate is usually made up of a number of costs. Certainly machine cost, rent or lease costs, utilities, and other overhead costs will be prorated out per machine. Often, labor costs for one operator and a portion of a material loader will be added. Sometimes this is added to the formula as a separate cost. As you can see, the machine hourly rate is only a significant figure if you know of what costs it consists.

Once the cost to mold the part is established, it is added to the material cost. To this is now added such costs as inserts and installation, bonding, machining, assembly, packaging, or whatever. These costs when totaled up make up the factory cost.

C. Additional Costs

Additional costs are now added, such as sales expense, commission, advertising expenses, and office overhead. A persent margin or profit is usually added to this. Some molders will add this into the machine rate.

D. Base Price

We now have established what is often called a base price. However, one other item which is totally volume sensitive has to be added; this is set-up.

E. Set-Up

In the structural foam business, this is probably the most misunderstood item of cost. Since you cannot prorate it over the price of a product without knowing the quantity to be run, set-up is usually added into the price as a volume-related add on.

The cost of setting up a large structural foam machine, especially a multinozzle one, can be very expensive and a major set-up could take 48 hr. Obviously the cost of this will be passed on to the customer. Material used for set-up is also a factor, and this cost must be included. Although the cost of tear down is not as high, it is still a factor to be added.

You can see how the cost of a low-volume run can soon build up. Most molders of structural foam hold to a minimum run of 250 parts due to this high set-up cost.

F. Review of Quotation

In reviewing any quotation there are several key items that must be defined before the quote can be considered valid. These items are:

1. Material type by grade
2. Cubic inches of the part (or weight)quoted
3. Tooling material of construction
4. Delivery time for both tool and first production
5. Terms
6. Volume break-down on set-up charge
7. Part number with revision level for which the quote was made
8. Piece price

VIII. ANALYZING SPREAD SHEET COSTS

Many purchasing people use a spread sheet when making a decision on which molder will produce their structural foam part. For picking a commodity item, this is no doubt a viable purchasing tool. However, for a plastic project having a wide variety of variables, the technique must be used with extreme caution. Items compared include: tooling charge, part cost, finishing cost (if applicable), delivery times, and terms. Usually a specific volume is picked and the piece price at that volume compared. Be careful of this since lower volumes tend to be heavily weighted with set-up charges. Volumes in the range of 1000 to 5000 are best to compare because they include set-up, but do not over emphasize it.

A. Tooling

In the tooling area, be sure you know for what your vendor is bidding. You cannot compare an aluminum tool with a steel tool anymore than you compare a pick-up and a jeep. They do different things for you and a judgement based on only "tooling" is not valid.

B. Materials

Also be sure materials are the same between bids. The difference between an impact polystyrene with flame retardant and a polycarbonate can cause a 50% differential.

C. Secondary Charges

Also be sure that the price includes the same secondary charges. Inserts, for instance, can add substantially to a part cost.

Most of the above can be settled with adequate engineering drawings that spell out exactly what you want. Over 20% of the time, the above "specifics" are left to the discretion of the molder to bid or not to bid. In some cases, even the material is left off. How can you compare apples to apples when each is picking their own fruit?

D. Trial Bid

A good technique, if you're looking for a good molder to work with, is to send out a trial bid with the above items called out.

Send it out to a number of molders and compare their bids as to completeness and clarity.

Once you have determined a good molder in this manner, bring him in and let him work on the program at the design stage. If he is worth his salt, he will direct you in the right direction, even if it means recommending another process. Most molders do not want a misapplication and will direct you accordingly.

A plastic project, to be successful, must be a joint program between the OEM designer/engineer, the moldmaker, and the molder. It is not something you can buy off the shelf based on the lowest bid.

E. Trust

Trust your molder or get a new one. Do not take the lowest bid or you will almost certainly end up with inflated engineering change costs, hidden extra costs, or substituted material. Most molders charge for material plus a machine rate. Material prices are relatively stable, and a molder's machine costs are nearly the same based on the economy and the labor area. Because all of these things are fairly close, what you are really looking for is service. A good custom molder is really your plastic consultant. How often do you take the lowest bid when hiring legal services or design consultation? The same should be true with your molder. Keep several with which to work, but pick the ones you can rely on to give you a straight answer.

IX. THE PROS OF PROTOTYPING [4]

Best said by designers themselves, the following came from *Plastics Design & Forum* magazine: Besides good design, prototyping is probably the best method of preventing costly and embarrassing product failures. Yet it is often skipped in the product development cycle. The stories of money and time wasted because of failure late in the product development cycle are all too numerous, even though rarely mentioned. Very often, the need for prototyping and product development service must be justified to a cost-conscious management who may see it as just a production aid.

The best argument of all in favor of prototyping is also the simplest one. Mistakes are made, and when they are, adjustments must follow. In the case of plastics product development, there is a time to make these mistakes. It occurs during product development. If the mistakes are not caught there, they are very expensive after tooling has been cut.

X. MODELS

Prior to finally building production tooling for a structural foam project, there is the decision of whether or not to build a model. This is an expensive decision and one must weigh all the advantages against the expenditure of both cost and time.

A model can be built to print usually using hard wood or plastic. Such a model is usually fabricated and certainly is never as strong as the final molded article. For checking the accuracy and detail of a print, there is no better way than building a model. Where parts need to mate with other parts, it is even more critical.

A. Timing

Usually a model can be produced during the time a purchase order for tooling has been let and the time the moldmaker starts cutting metal. Often the moldmaker himself will build the model. During this period of model making, the toolmaker can lay out the tooling design, purchase materials, and even start on some components of the tool without needing the final go ahead to start cutting cavities and cores.

In every case, where a model was built on a project, it has shown areas where further design information was needed. It is not to say that these areas might not have been correctable in the hard tooling, but it certainly was less expensive to build the model.

B. Inside Model

One technique that often can be used is to build only an interior model. Whereas the outside is rather straightforward and the inside is supercritical from a placement standpoint, an inside model will show quickly the accuracy of the prints.

C. Exterior Model

When a customer is really not sure what the part should look like, a model of the exterior only is used. This is certainly less expensive than a full model and can be used for getting agreement on aesthetics as well as for advertising purposes.

Structural foam applications with optimum 1/4-in. walls are fairly easy to model. Hard wood or foamed sheet stock is usually available in 1/4-in. stock. With most structural foam sheet, solvent bonding can be used to fasten walls together. The cellular area in the center often shows up when building a model in this way and it is difficult to cover up.

However, models built in this manner will often give valuable information when used to run heat build-up tests. When this type of information is required before final design, this approach offers a quick and money-saving technique.

Material suppliers to the structural foam industry often have sheet stock available at little if any cost.

XI. ENGINEERING DRAWINGS

Most molders of structural foam build their molds to produce parts to print. It is therefore critical that the drawings provided with the purchase order are exactly what should be purchased. Any unanswered questions or verbal answers to questions, or interpretations, are problems ready to happen. Some can be very costly. This is, of course, where a to-print model comes in handy.

A custom molder who is doing his job will call to your attention any problem areas he finds. If he sees a need for another cross-section or a detail of some difficult-to-interpret area, he should ask for it before the final price is settled.

A. Dimensional Tolerances

Dimensional tolerances are always an area of concern. Because of the nature of structural foam, shrinkage in one part of the mold may not be the same as in another area of the mold. All dimensions on a part print are not going to be met exactly in the molded part. In most cases, dimensions are put on a part because it has to have some dimension. Critical dimensions should be noted before building the tool. It is often more important that "part A" fits together with "part B." The actual dimension may be superfluous.

Holding a molder to tighter tolerances than is actually needed will increase costs. A scale for tolerances on certain materials is included below. However, remember that by opening tolerances on dimensions that are not critical can save both time and money. (See App., Sec. VIII.)

1. Filled Products

Filled products such as 20% glass-filled polycarbonate shrink very little in the mold, even in the foamed form, and consequently you can hold extremely tight dimensions. Polyethylene or polypropylene, on the other hand, have high shrinkages, so tolerances must be loose. Material selection can affect the ability to hold tolerances dramatically.

B. Extra Print Copies

Faster, more accurate quotations on structural foam parts will be provided if at least three copies of the drawings are sent. With all the duplicating equipment available today, one should be able to supply at least a reduced set of drawings. This is a godsend to a molder's estimator for it allows him to send copies to the moldmaker, the finisher, and to have a set available for himself. Your quotation will be more accurate, returned quicker, and might even be lower.

XII. THE NEED TO KNOW

As we have explained in the previous chapter, there are several ways to produce structural foam. The question can be asked, how much of this information is needed to design a structural foam part? How much is needed to purchase a structural foam part once it is designed?

Certainly the designer or engineer who has been given the job of designing a part should know, once he has decided on the parameters of a part, whether it can be produced.

A. Wall Sections

Wall section seems to be the major area of concern with a foam part. For many years, the optimum wall thickness was 1/4 in. This provided for a relatively smooth surface, a very strong structural foam part, and, of course, low pressure in the cavity during molding. By keeping to a 1/4-in. wall, the cycle, while still long when compared with solid injection molding, will be $2\frac{1}{2}$ to 3 minutes overall. With multinozzle machines, whole sets of parts or multiple cavities make this cycle reasonable because you are getting several parts with each shot.

B. Surface Definition

As finishing became more critical and OEM demanded less costs in preparing this surface, thinner walls and high pressures were introduced to give better surface definition and less sanding. This approach did help, but it also increased the molding pressures and made low-cost aluminum tooling a risk. Parting lines were more vulnerable and tool life was reduced. The answer was, of course, more expensive steel tooling and we are back to an almost solid, barely foamed part with thin walls and a shorter cycle time.

Over the years, dozens of techniques and processes have been developed to accomplish the same result. Some are lab curiosities, some are commercial realities. Tooling varies drastically from process to process and, of course, so do the results. It is no wonder that the designer and the purchasing agent have become confused. Even the industry cannot make up its mind which technique is the best.

C. Define the Product

The best approach is to define the product you wish to build, spell out the parameters, and get a quote on each process that seems to apply. Economics have a way of separating the right way from the wrong.

Many things are possible if money is no object. However, few OEM have that luxury. Some people like to be the first using a new process or technique. However, pioneering can be very expensive and may, in fact, jeopardize the whole project.

Dual injection, expansion web molding, and other procedures are all potential methods to produce a part. They are also mostly untried or inherently higher in cost. This is not to say that they might not do the job you require. But look at the conventional methods first, then if you are not satisfied with what you see, go to newer methods. Be sure you don't pay for research and development on a molding process without some reasonable assurance of success.

XIII. WORKING WITH A CUSTOM MOLDER

Over the years, a technique of purchasing structural foam parts, especially for the electronics industries, has evolved. This approach has caused many custom molders to go out of business and certainly has not done the cost-conscious electronic industry any good.

A. Price Alone

By purchasing on price alone and by doing so even before final design parameters have been set, the industry has made it almost impossible for a molder to make a decent margin on his work. The industry has demanded more and more, from zero defects to limited inventory shipping techniques, without feeling the necessity to pay extra for the extra service. If a molder goes out of business, too bad. There are always others ready to pick up the purchase orders.

On the surface, this looks like a win-win situation for the electronic OEM. He has a continuing source of low-cost housings to put around his electronic package.

B. Hidden Costs

However, the hidden cost of changing suppliers is borne by the OEM. In times of excess capacity in the molding community, the OEM can do this without fear of cutting himself off from his supply. However, as happens on an almost cyclic basis, as business expands, the molding machines of most molders become full. The more experienced molders will always fill their equipment first. Now the constant pinching of price will come back to haunt the electronic OEM.

C. Vendor Loyalty

A much better approach that develops a molder's loyalty is the early establishment of a working relationship with the molder. As the project develops, both you and the molder have a desire and a need to bring the project to completion in the shortest time and at a fair price. The results, especially in the cutting of time, will be well worth the effort and you will also have established a working relationship for the next project.

The automotive industry, a long-time abuser of small molders, has seen the value of a close relationship.

A typical situation reported by *Plastics Focus* gave the following report:

> ". . . what seemed to interest the audience even more than the technology was the way that Pontiac did the sourcing for the car. Vendors were chosen not on a cost basis (no quotations were asked for, or expected), but on the basis of technical expertise and manufacturing capabilities. Early sourcing—even before the design was put to paper—was also practiced, so the vendors (chosen by visiting teams of Pontiac purchasing agents, technicians, etc.) could actually help in the planning stages [5].

The electronic industry could benefit by adopting this technique.

REFERENCES

1. Jordan Rotheiser and Glenn Beall, Designing Products That Sell, *Plastics Design and Forum*, Industry Media, May/June, p. 13.

2. Milton Montz, *The Handbook of Structural Foam - A Business Approach*, internal document, Beckman Instruments, Irvine, California, p. 24.

3. Bruce C. Wendle, *Molder Selection Criteria*, 1979 Conference Procedings, S.P.I. Structural Foam Division, Technomic Publishing, Inc., 1979, p. 1.

4. The Pros of Prototyping, *Plastics Design and Forum*, Industry Media, July/August 1976.

5. Joel Frados, *Plastics Focus*, Plastics Focus Publishing, Inc., 15:3, No. 32, Oct. 3, 1983.

6

Secondary Operations

I. INTRODUCTION

This chapter relates to the general feasibility and purchase of secondary operations performed on structural foam. These operations can often cost as much as the molded part itself and must be screened carefully when reviewing the cost of a product.

II. FINISHING

The cost of finishing a structural foam part is probably the most talked about and certainly the most misunderstood part of any quotation. The need to have a product look good is really only part of the story. By applying a coating of two-component polyurethane paints onto the foamed part, one is accomplishing a great deal more than just making it look nice. The urethane layer, which attaches itself chemically to the foamed plastic part, provides excellent abrasion resistance, much improved chemical resistance, and long-lasting color integrity. All of these features are needed improvements on a noncoated plastic whether it is foam molded or injection molded.

This type of a system (urethane over a foamed part) provides a molded part with exceptionally rugged, long-lasting quality with very few other competitors. The larger the part, the more economical this system becomes.

A. Purchasing a Finished Structural Foam Part

When purchasing a structural foam product that is to be finished, one must have a good understanding of what is involved. Because of the swirl surface on most foam parts, the need to cover the surface of the visual portion of the part is common.

Each application will vary in just how a part is to be covered and there are as many variations to this as there are products. Some products, such as waste receptacles and rubbish cans, need no coating. Here the colored plastic, usually high-density polyethylene, stands up well to hard abuse, the color is molded completely through the part and the swirl is actually quite acceptable. (See App., Sec. II.)

B. One-Part Color Coat

On some parts where an inexpensive color coat is needed, one might use a water-based acrylic. This type of finish used directly over an as-molded surface looks pretty decent and adhesion to most of the styrenic based materials is quite good.

The swirl will normally show through. But because there is good mechanical adhesion, the part will stand up to some abuse. A dark surface on the inside of an optical instrument would be an example of this type of finish.

C. Two-Component Coatings

For best all-round exterior coatings, for such products as instrument housings and computer peripheral enclosures, the two-component polyurethane is most often specified. More than just a layer of paint over a substrate, this reactive coating actually bonds itself to the structural foam part and becomes a part of the system that is second to none when it comes to abrasion resistance and general all-around toughness. This system is far superior to the plastic surface itself. The soft thermoplastic is subject to scratches and abrasions as well as being susceptible to solvents, stains, and other chemicals.

D. Surface Preparation

The high cost of labor needed to prepare the molded product for painting is what has caused many OEM to look for other ways of manufacturing a product. The paint itself is usually the same type as that used for fabricated and cast metal, but surface preparation can make it more expensive.

Also the fact that most plastics cannot take the high bake temperatures used when coating metal makes the surface preparation more critical. A part is finished after molding as in Fig. 6.1.

Even with the textured surfaces normally used on such items as business machine housings, the surface underneath must be prepared properly for finishing. This might include sanding before application of base coat to remove parting and witness lines. A light sanding is almost always required between base coat and color coat.

Figure 6.1 The painting of structural foam parts has become one of the largest post-molding costs required. Most products get a two- or three-coat coverage of two-component urethane which provides an exceptionally tough chemically resistant finish.

E. Base Coat

Standard procedure calls for a base coat. This is then lightly sanded and the color coat of two-component polyurethane applied. A system such as Sherwin William's Polane T is typical. A spatter coat is then applied over the color coat to provide the texture.

When deep swirl, often called "elephant skin," is present, additional sanding is required both before and after the base coat. The design of the mold used to produce the structural foam part can have a great deal to do with the formation of this heavy swirl. Uniform wall sections, usually in the 1/4 in. range, will also keep this effect to a minimum.

On any given part, a weight standard heavy evough to fill the part to adequate density will help to reduce this heavy swirl.

Prices on finishing are normally based on so much per square foot of applied surface. However, any masking that needs to be done will often increase the price dramatically.

F. Two-Color Finishing

Two-color parts are by far the most expensive and it is often worthwhile to look at molding the two-color part in two pieces, painting each separately and bonding back together.

If you must have two-color parts, use a "V" groove between colors so that a metal mask can be used.

G. Quality Control

In the quality control of painted parts, there are really two problem areas: texture and color. Both are highly susceptible to human error and need to be monitored continuously. A color chip system produced from the same color batch should be prepared with both correct color and texture. Signed and dated samples should be supplied to both the molders' quality control and your own quality control people. All others should be destroyed. A specific numbering system should also be developed.

No changes or exceptions should be allowed without going through purchasing and then only by going through the same color chip approval procedure. The test light sources used to view these samples should also be determined and specified. Texture and color both can look much differently under two different light sources. However, all of these preparations will not eliminate all the problems. Since the spatter or texture is still put on by a painter with a spray gun, getting exactly the same finish consistently is difficult.

As with all such purchases, the need for close control should be compared with the added cost. If an instrument seldom ends up next to another of the same manufacturer, it hardly seems worth paying extra money to assure that the two coatings are identical.

Further comment will be made of conductive coatings later, but it is important to note that this new requirement for shielded enclosures will also increase your costs dramatically.

H. Electrostatic Finishing

A new technique using electroless plating of both copper and nickel as a shielding system also offers the possibility of electrostatically spraying a structural foam part. This reduces overspray, gives a more uniform coating, and cuts costs.

III. COOPERATION BETWEEN SUPPLIERS

Working with your molder and finisher early in the design stages can often save you many dollars and valuable time. If expensive finishing requirements are not discussed before designing the part, you may find a good share of your budget being used for cosmetic finishing. Today as much as 30% of the cost of a part can be tied up in finishing.

Some companies request precolored material to be used in parts which are then painted with the same color paint. This is done to eliminate show or wearthrough in high-wear areas. However, unless you also color the base coat you will have a layer of another color, usually white, between the colored plastic and the urethane. Wear-through on business machine housing has never been a major problem and it is doubtful whether this added feature is justified.

To help reduce finishing costs, one should be on the lookout for ways to eliminate them. Parts such as the base of an instrument, which has only a minimum amount of area showing, could be molded in the proper color and not finished. Instruments needing a black or dark interior can be molded in the right color with no additional interior finishing needed.

Large flat areas on a part might be decorated less expensively by utilizing a large pressure-sensitive decal or by replacing the molded surface with a decorative metal or wood panel. Obviously functional requirements must be met first. However, once they have been addressed, decorative techniques of many kinds are available to the manufacturer. Again, let your molder, your finisher, and your industrial designer help you find the least expensive way to protect the item.

IV. VAPOR POLISHING OF STRUCTURAL FOAM PARTS [1]

A new technology that can provide an injectionlike finish on a structural foam part for a relatively low cost is called vapor polishing. Adapted from the technology used to polish polycarbonate eyeglass lenses, vapor polishing is a relatively simple technique for low-pressure structural foam parts that achieves an aesthetic surface appearance at a much lower cost than painting.

This procedure involves immersing a structural foam part in solvent vapors for 5–10 sec. These vapors dissolve and reflow the exposed surface, resulting in gloss replacing the swirled, dulled surface of the part. Optimum immersion time, depending on material, color, density, and texture, ranges from 5 to 10 sec. This eliminates both the swirl and pinholing.

Preliminary test results indicate that structural foam parts with low-density reduction are best suited for vapor polishing, and that both density and immersion will affect the level of gloss. The development of this one-step finishing operation is particularly timely, given the increased competitiveness of the marketplace. Vapor polishing may never replace painting, but the preliminary successes show it to be a major step towards helping to reduce structural foam part costs.

V. EMBOSSED PART SURFACES

Another technique being used to break up the swirl surface of structural foam is to use mold embossing. All types of patterns and surfaces are available and competition has made this technique less expensive.

In foam parts where internal pressure is low, the surface definition is not as good as on standard injection parts. Because of this, one should look at a fairly heavy embossed surface—at least 5–6 mil into the mold surface. This will assure you of getting a defined pattern in the molded part.

This may also present a possible design problem. Embossing on side walls tends to build undercuts in these areas, keeping the part securely fastened to the cavity. The rule of thumb is that a degree of draft is needed for every mil of depth of embossing. This means that your part must have 5 to 6° on the vertical wall—if you are going to have an effective embossed wall surface.

In thin-wall foams and foams molded on high-pressure machines, embossed surfaces can be covered with just one coat of urethane paint, allowing the embossed surface to provide the pattern.

On a new mold, be sure the part is sampled first before any embossing is done. Once the embossing is put into the mold, it is almost impossible to repair or change an area within the mold without visably altering the embossed pattern.

VI. SHIELDING OF STRUCTURAL FOAM

Electromagnetic compatibility (EMC) may be a new term to many plastics designers, but with the increased selection of structural foam and other plastics for electronic equipment enclosure applications, EMC is making its presence felt in a major way. In a worst-case situation for the plastics industry, failure to meet EMC constraints could force enclosure redesign back to metal. (See Fig. 6.2.) The ability of electrical devices to function normally without being interfered with, or without interfering with other electrical devices, is what is thought of as electromagnetic compatibility.

Electrostatic discharge (ESD) affects the plastic industry in much the same way that electromagnetic interference (EMI) does. Plastics are generally excellent insulators and therefore do not allow charges to bleed off to ground in a controlled manner. Instead, discharges to the plastic case, in many instances, can cause equipment malfunction or failure [2].

A. Electromagnetic Interference

EMI testing must be done on the materials under consideration for use as a shielding approach as well as on the individual electronic equipment undergoing design. The shielding effectiveness figures, usually given to the industry by materials manufacturers, are provided as a guide to show the relative level of shielding performance on an ideal enclosure. That enclosure is one which would be solid, without holes, for display, keyboard, access, ventilation, power, or seams, etc. This is not a practical case, however.

Properly designed vents must appear to be electrically solid and input/output cables need to be appropriately grounded and shielded to minimize their degradation of the equipment's EMI signature when it undergoes EMC acceptance testing.

FIGURE 6.2 Shielding of structural foam parts has become a necessity when they are utilized as enclosures for the electronic industry. Here an arc spray of zinc is applied to the inside of a printer housing.

B. EMI and ESD

Consideration of the EMC requirements must address EMI and ESD design goals in shielding selection. If ESD is the major concern, too conductive a shield may give rise to continued EMI problems [2].

C. Shielding Techniques [3]

It is now quite apparent that a significant challenge exists in EMI shielding when plastic enclosures are used in various electronic applications. Nonetheless, the use of plastics is proliferating for the following reasons: design freedom, cost and weight reduction, and reduced manufacturing problems where units of complex geometry are required. These advantages over metallic components appear to outweigh the disadvantage of having to add a conductive barrier to the plastic substrates.

To apply the conductive barrier, a variety of approaches have been considered. The key to the selection of the optimum method lies in evaluating the cost/performance differences between methods. Selection criteria generally vary according to desired physical characteristics, such as frequency, field strength, production volume, appearance, cost, component sensitivity, and use environment. Whatever the method chosen, it must be a readily controllable, cost-effective technique that consistently provides a highly conductive coating of uniform thickness.

Until now, these criteria have been approximated but never fully achieved. Although some of the conventional techniques can be made to provide adequate shielding, numerous performance and cost deficiencies diminish their appeal to many end users.

The following outline identifies the primary deficiencies of current shielding technologies.

Flame or Arc-Sprayed Zinc

This system utilizes the atomization of zinc wire by either a gas flame or an electrical arc to apply a coating of zinc on the inside of the plastic enclosure. Shielding characteristics of this system are excellent.

Uniform spray application is often obstructed by bosses, ribs, and corners. Undesirable dimensional build-up may also occur. Powdering, flaking, and chipping are characteristics of spray-applied coatings. A grit-blast surface preparation is required to augment adhesion. Health hazards of grit-blasting and airborne zinc require OSHA compliance. Waste and unnecessary build-up due to overspray or excess spray during application results in appreciable costs.

Metal-Filled Paints

At present the most widely used technique is the application of metal- (usually nickel) filled acrylic paint to the surface of the housing. The process is subject to spray application inconsistencies similar to those of sprayed zinc.

Constant agitation is usually required, due to rapid settling of metallics in the binder. Multiple coats are usually required to achieve acceptable attenuation, and this increases material and labor costs.

Conductive Fillers

Certain conductive fillers such as carbon and metal fibers can be put directly into the plastic resin prior to molding. Resin displacement tends to weaken the structural integrity of the plastic. Conductivity depends on the random cross-linking of filler material. Woven and nonwoven conductive mats tend to flow inconsistently. Increased molding complexity may preclude common injection techniques.

Vacuum Metallizing and Cathode Sputtering

Both of these techniques use the vaporizing of a metallic element, usually aluminum, which is then applied in a thin layer to the plastic surface. The equipment needed, usually a vacuum chamber, is expensive and has size limitations when processing large structural foam parts.

Foil Applications

Thin metal foil is cut and formed to cover the potential emmission source which increases cost.

Silver Reduction

A layer of silver is deposited on the plastic surface by means of a liquid reduction system. Masking is difficult to achieve. Raw material cost alone is significantly high.

Electroless Plating

This process involves the electroless plating of metal, usually copper or nickel, onto the surface of the plastic part. A coating of a conductive metal in solution is applied first by dipping the plastic part into a large tank. This is next put into another bath where the first metal ion is replaced with copper or nickel ions providing the conductive surface.

The system is not unlike the conventional electroless plating systems used for years to plate plastics with high-gloss chrome surfaces. Some problems with adhesion and conductivity developed when platers tried the conventional systems as a conductive coating for shield.

A research effort initiated by Occidental Chemical Co. was directed toward resolving these problems. The result of this development was ATTENUPLATE,* an electroless (nonelectrolytic) plating system custom-tailored to meet EMI shielding requirements. The features of this shielding system provide significant advantages over current shielding approaches.

Electroless Deposition from an Aqueous Bath: Complete coverage, with consistent uniformity of thickness over the entire plastic surface, is obtained, regardless of the physical geometry of the part. A continuous deposit is produced that does not flake, chip, or powder. Cost efficiencies are realized because there is little or no metal waste, no organic binder, and no electrical application requirement. Further efficiencies are achieved through bulk processing.

Variety of Coatings: Attenuation-specific (custom-tailored) shielding can be achieved simply by varying the coating thickness and materials used. Optional top coats over electroless nickel or copper base coats provide the specific surface properties desired.

ATTENUPLATE products may be further enhanced by using electrolytic coatings for ultra-high-performance shielding and certain decorative effects. Stop-offs may be employed for selective area plating. Both plated and stopped-off surfaces accept paint readily.

Substrate Compatibility: The ATTENUPLATE system is compatible with the most widely used substrate materials. Foamed and injection-molded ABS and PPO are well suited and current development work is addressing compatibility with a variety of other promising materials. Physical characteristics of flame-retarded PPO and ABS, both foamed and injection-molded, are not degraded when coated with ATTENUPLATE deposits.

---

*Trademark of Occidental Chemical Corp.

Commercial Availability: Production-proven process technology is available. Unused electroless plating capacity already exists nationwide in job shops and captive houses to accommodate EMI shielding requirements.

Performance data (under worst-case conditions) indicate that EMI shielding attenuation exceeding 45 dB (frequency range of 10 MHz to 1000 MHz) can be achieved using such systems in thicknesses as low as 30 microinches [3].

Because this system puts a metallic skin on both sides of the part, it lends itself for use on structural foam. Adhesion of the urethane coating to the plating is excellent and the resulting system provides an exceptionally well-shielded housing material.

Design Solutions

Coatings or metallization methods are only part of the EMC solution. The rest of the solution lies in circuit design, layout, and conventional grounding and filtering considerations that have been left until the system's final days of design—or worse, until after design finalization has taken place and problems occur.

VII. ASSEMBLY METHODS

There are many ways in which to assemble structural foam parts, including mechanical fasteners such as screws and inserts; adhesives and solvent bonding; ultrasonic and induction welding; and the use of hinges and snapfits.

A. Adhesives

Epoxies, urethanes, acrylics, cycroacrylate anaerobics, and silicones are available for use on some structural foam products. Adhesives can offer labor, weight, and material reductions, if used properly.

B. Solvent Bonding

This is similar to adhesive bonding. The outer layers of the plastic are softened by the solvent, clamped or pressed together, and are bonded together when the solvent evaporates. Base resin may be added to the solvent to create a slurry solution and possibly a more effective bond. This method is simple and inexpensive.

C. Welding—Ultrasonic

Sonic pulses are transmitted to the part by an ultrasonic horn causing two plastic materials to vibrate, heat, and then fuse together. This process is quick and clean. Materials to be bonded together require similar melt temperatures.

D. Welding—Induction

A metal wire, insert, or powder is placed into a plastic joint, held tightly together, and heated using a high-frequency magnetic field. The combination of compression and heat produces an acceptable fusion weld. Induction welding is a more expensive technique often used for difficult-to-bond plastics and irregularly contoured surfaces.

E. Hinges

Integral hinges can also be used in foam if polypropylene is the base material. Other materials will provide limited hinge life, but a polypropylene hinge, properly stressed as it comes from the mold and before it cools, can yield thousands of cycles.

All of the above mentioned techniques should be reviewed in the early design stages to see what the most economical technique might be.

F. Jig and Drill Fixtures

Once the design is established, it is important to bring the molder or the people who will be doing the assembly into the picture. Properly designed jig fixtures are part of the cost of producing the product and should be part of the quotation package—not an afterthought that throws the budget out of line at the last minute. For applications that will be assembled more than once, a designer should consider the following:

G. Metal Inserts

There is a special form of nut that acts as a tapped hole. Parts using such inserts can be reassembled many times. They are installed using the following techniques:

Ultrasonic

Ultrasonic vibration causes frictional heat between the metal insert and the plastic, and this melts a thin layer of plastic, which solidifies around the undercuts of the insert. One should follow the design recommendations offered by the fastener supplier when designing holes and busses for these inserts. A typical recommendation is included below.

Expansion

Expansion or pressed-in inserts have locking mechanisms, knurls, or retention rings that engate the base material as they are installed. They are easy to install and are used for lightly loaded applications.

H. Screws

Thread-forming screws form a mating thread when driven into molded preformed holes. These screws allow for rapid installation and can be disassembled between 1 and 20 times. In foam, a hole can be drilled rather than molded in, but the holding power is not as high due to the cellular core into which it goes.

I. Molded-In Hinges and Snapfits

Hinges and snapfits can be designed directly into the part and can often result in substantial labor savings. A part may be disassembled many times using these techniques.

J. Overdesign

The use of metal (usually brass) inserts to hold parts together or to secure inside components to structural foam parts is widespread. Most structural foam molders install literally thousands of these, usually by ultrasonics and as a postmolding operation. Automated systems for installation can reduce costs of these items, but they still add extensive costs to a molded part.

The electronic industry has, over the years, gotten into the habit of putting metal inserts into every hole in the part, adding greatly to the cost. In most cases, if the fastener being fastened into the insert is not taken out more than once or twice in the lifetime of the product, one can get excellent holding characteristics by inserting the fastener directly into a cored hold in the plastic.

The characteristics of thread-forming screw make it ideal for use in the compressable foam structure and the metal insert just is not needed. However, if the screw or fastener is going to be taken out and put back in many times, the metal insert is probably a good investment.

K. Molded-In Inserts

Some designs require the molding of metal inserts directly into the foam part. This can be done and does offer some design possibilities. However, there are problem areas to watch out for. Sharp corners on molded-in inserts can and often do cause stress-cracking. This is very true in solid injection-molded parts and to a lesser degree in the foam. Also the difference in shrinkage between the metal and plastic can also cause stress-cracking. The plastic must be allowed to move as it cools, otherwise stress will build up and stress-cracking occurs.

One approach that can often be used is really a postmolding technique. If you insert a cool metal insert into a carefully designed hole as the plastic part comes from the machine, the material will tend to shrink around the metal insert as it cools and hold the insert securely. This is good for bearings or shafts that you may want tightly held in a structural foam member.

Urethane foams lend themselves to molded-in inserts due to the fact that they are low-viscosity liquids when injected into the mold. As they foam in the mold, less stress is generated as they expand around the insert.

VIII. MACHINING

Structural foam parts can be machined using standard plastic machining techniques. However, the foamed core causes some problems.

Sheet stock of several foamed materials is available from suppliers and models can be fabricated. These are often used for heat testing and strength tests. However, one should remember that the foam gets its strength from the integral skins. Once these skins have been cut through, they will no longer be as strong as the molded item.

IX. LABELING [4]

Labels on structural foam parts are not only a good way to decorate parts, but they also can be used to hide blemishes in the surface, such as the gate where material was entered. New techniques in label manufacturing, as well as new bonding methods, have made this a valuable decorating tool. All types of substrates are available from which to produce labels.

A. Fastening Methods

The method used to fasten the name plate to a product depends on the product and the surface to which it is applied.

Metal Fasteners

Metal fasteners include sheet metal screws, drive screws, escutcheon pins, bolts, machine screws and rivets. All can be used with structural foam.

Adhesives

The use of adhesives to hold a name plate in place eliminates the necessity for other hold-down methods such as screws, bolts, rivets, etc. Due to advancing adhesive technology, there is an ever increasing usage of name plates and decorative trim on end products. Cooperative engineering, involving the end user, the name plate manufacturer, and the adhesive supplier, have made this a very effective method.

Major advantages obtained from adhesive bonding include: lower attachments costs; greater flexibility in design of name plates, decorative trim and the end products; aesthetic value of a neat assembly; and a satisfactory, permanent attachment of the name plate or decorative trim.

The adhesive selected for a particular application should be one with an immediate working bond that is sufficient to hold on until the unit is packed, and then will develop the strength necessary to resist the stress and environmental conditions to which the bond will be exposed. Tests should be delayed for the specified interval after bonding.

To obtain the best adhesive for a particular application, the end user should decide what basic type of adhesive will best fit into the manufacturing process and then design into both the name plate and his product the conditions necessary to accommodate the adhesive.

To assist the name plate manufacturer in suggesting and furnishing the best adhesive, every blueprint for an adhesive-backed name plate should include information regarding the part to which it is to be attached. Along with the chemical identity of the surface, facts concerning the degree of flatness, texture, shape, etc., should be included. Direct reference should be made if the bond is to be expected to withstand specific environmental or stress conditions.

Adhesive manufacturers present basic information regarding the adhesion to various surfaces, resistances to environmental conditions of design, and application that will contribute to the attainment of an economical, trouble-free, permanent attachment. Cooperative usage of the information will enable the end user and the name plate manufacturer to identify readily and choose a most practical adhesive for any particular application.

All adhesives, whether of the pressure-sensitive or activated type, depend on a mechanical and/or chemical action to adhere. Tackiness alone is often a misleading factor in judging an adhesive.

B. Design Technique

Part Configurations

Parts can be of any design, size, or shape. The only limiting factor is adhesive contact to the substrate.

Substrates

The substrate onto which the name plate will be applied should be designed with the following items in mind:

Tolerances. The area is large enough to insure that the name plate will fit, considering the maximum tolerances of the name plate.

Thermal expansion. The area is large enough to compensate for differences in thermoexpansion of the name plate and/or substrate.

Recessed or inlays. Recessed areas usually do not have sharp inside dimensions. This fact should be kept in mind when designing and specifying the dimensions of name plates to be inlaid.

Large Parts

Large name plates and rigid parts do not conform well to the substrate. This can hinder the complete adhesive-to-surface contact necessary for optimum bond strength.

Metal name plates tend to be slightly distorted. This distortion is evident if the part does not lay down flat on the substrate. The memory in the metal can cause the part to lift. Precurving the part or using a thicker adhesive system should compensate for this problem.

Rigid parts do not conform to the substrate. Therefore, be sure the part lays completely flat on the substrate. It is possible to compensate for irregular adhesive-to-surface contact with a thicker adhesive system.

Gripping, Not Lifting

Edge Lifting: Edge lifting can be caused by the name plate or panel recovery at the edges. The recovery is due to the elastic memory of the panel exceeding the adhesive's holding properties. This phenomenon can be accelerated by exposure to hot or cold environments.

Curved Substrates and Name Plates: Curved substrates require the panels to be preformed prior to application, otherwise the lifting forces concentrated at the edges will cause lifting. The panel should be preformed equal to or even less than the curvature of the substrate.

Preforming: Applying preformed name plates to a flat substrate to obtain the gripping action is common practice. The amount of preforming should not be great. A successful law of curvature is 1/4 in. per 4 in. of length. The amount of curvature will vary with the rigidity of the panels or nameplates. A rigid 20 mil full hard aluminum panel 4 in. long can obviously not be curved 1/4 in. per 4 inches and still lay down satisfactorily. A simple test for evaluating the degree of curvature should be conducted prior to preforming of the parts.

Painted Surfaces: The bond of the paint to substrate is vital as this would have direct effect of adhesive bond to the paint. This is especially the case with structural foam.

C. Pressure-Sensitive Type

This type of adhesive contains chemical components that cause the adhesive to have a tacky feel at all times and so may be easily applied by the addition of pressure. Due to its constant tackiness, the adhesive is protected prior to application by a paper treated with a release coating. This paper is usually referred to as a liner. Most name plate manufacturers buy their pressure-sensitive adhesives in tape form and apply it on sheets of name plates by pressure through pinch rolls.

To apply, the customer must peel the liner to expose the adhesive, being careful not to touch or contaminate the exposed adhesive. There are some devices available to aid in the removal of the liner. The surface to which the name plate is applied must be clean and free of dust. The name plate is then pressed down either by fingers, roller, or platen.

D. Solvent-Activated Types

This type of adhesive is a dry film and has little or no tack before solvents are applied. You may purchase name plates with this type of adhesive that do not have a release-coated paper for protection (dry back), or with a liner to protect it from contamination prior to application.

E. Heat-Activated Type

In recent years, this type of adhesive has gained in popularity. It consists of a dry film similar to the solvent-activated type. Most solvent-activated types may also be heat activated. Nearly all heat-activated adhesives are of a thermoplastic material. Thermosetting adhesives are usually not used by name plate manufacturers.

F. Solutions to Improper Applications

1. Contaminated surfaces can cause poor adhesion. These conditions can be due to the following:
   a. Oil, grease, dirt, or mold release left on the surface.
   b. Silicone or wax in the paint finish.

   Surfaces can be cleaned with a solvent but care should be taken that the solvent used is compatible with the plastic polymer being used.
2. Some adhesives will not have an affinity for certain polymers. Polyolefins such as polypropylene or polyethylene require special adhesive formulations.
3. Improper or contaminated solvent used with solvent-activated adhesive. Under- or overactivation can cause poor adhesion.
4. Insufficient pressure applied during application of either pressure-sensitive or solvent-activated adhesives.
5. Check surface preparation by operator.
6. Check applying or bonding techniques (make sure of 100% contact).
7. Check if applying environment is unchanged (e.g., from air-conditioned room to humid area).

G. Suggested Test Procedures

In addition to environmental tests such as humidity, outdoor exposure, solvents, heat, chemicals, etc., the other usual test is for peel strength. The peel strength should be noted in pounds per inch of width. The thickness of the material should also be noted. Always advice name plate manufacturers to which the adhesive would be subject [4].

X. PACKAGING

The packaging of structural foam can also be a major cost factor and should be thoroughly spelled out before a molding contract is let. Most molders include in their price a packaging cost sufficient to box the item in bulk and get is safely to the OEM. A variety of methods are used for this and it will vary greatly between molders.

A. Packaging a Painted Part

Packaging a painted housing ready for the production line is going to be considerably different than packaging a nonpainted object. Nesting is important as is the weight of the filled box.

Structural foam is pretty tough stuff and as such can be shipped easily without fear of breakage. However, the painted part, especially if the paint is still soft, is another matter. One method is to require a wrapping of 1/4-in. polyethylene or polyurethane foam around each item. This is usually taped in place and then placed into a predesigned box. The box should be filled in all areas. Large voids will collapse causing boxes to fall as well as damaging the contents.

B. Palletizing

Palletizing is a good way to protect the shipment, but this is added cost and handling.

C. Scrap Removal

In recent years, the cost of getting rid of the packaging materials used to ship product into the OEM has become a major problem. One way around this is to have the molder supply his finished housing to you in a package suitable for shipping your final product to your customers. This solves the packaging scrap problem and also assures you of getting your painted enclosure in house in first-class shape.

D. Expanded Polystyrene Cocoons

The use of an expanded polystyrene (EPS) cocoons or end caps designed for final product shipment can be used on the enclosure for shipment from your molder.

E. Markings

Markings on boxes should also be spelled out in the specifications. Part name, part number, purchase order number, and quantity in the box should be included in any marking system and labels or stencilling should be on at least three sides of the box.

XI. BREAKAGE AND REPAIR

If you do receive a molded product that is damaged, you may want to salvage it yourself. Welding or bonding of broken foam parts is possible and paint touch-up can sometimes be an easy cure. Obviously the molder should be brought in to discuss such problems and to share the costs if they are of his doing.

A discussion of packaging and handling of plastic parts prior to signing a purchase order will shortstop many problems that you may encounter later as the product is molded and shipped.

XII. FREIGHT COSTS

Freight costs for structural foam products are a big factor due to both size and weight. However, do not just look at the distance from the molder and assume the cost to be relative. Some areas have more backhauling than others and deals can be made, especially on truck-load lots. Piggy backing on freight trains can also be an effective way to cut costs.

REFERENCES

1. Stewart R. Levy, *Vapor Polishing of Structural Foam*, 1983 Conference Proceedings, S.P.I. Structural Foam Division, 1983, p. 126.

2. J. J. Coniglio, *EMC: A Problem Whose Time Has Come*, 1979 Conference Proceedings, S.P.I. Structural Foam Division, Technomic Publishing Co., 1979, p. 24.

3. Daniel J. Oberle, *Electromagnetic Interference Shielding Through Plating on Plastics*, 1983 Conference Proceedings, S.P.I. Structural Foam Division, 1983, p. 00.

4. *Name Plate Industry Standards and Practices*, National Association of Name Plate Mfgs., Inc., 100 Vermont Ave., N.W., Washington D.C., 20005, p. 22.

7

After the Loving

I. INTRODUCTION

During the time your new structural foam parts are being quoted, you have the feeling that things are happening. Molders and their reps are constantly calling, quotations are coming in, decisions being made. What about after the contract is let? What happens then? Do you go back to your office and wait the 12–14 long weeks to hear from the molder? How can you utilize this long mold-building period to the most advantage?

II. MOLD-BUILDING SCHEDULE

First be sure your molder provides you with a mold-building schedule. At what time in the production schedule does the toolmaker have tool design complete? When is the material for the mold going to be all in house? Ask for a weekly report on this and have your molder be sure the information he's passing on to you is accurate (see Table 7.1).

III. LAST-MINUTE CHANGES

Also of great concern are engineering changes made after the tooling purchase order is let. This is probably the biggest money and time consumer that can plague a purchasing agent. To keep on top of this situation you must do a number of things. First, find out from the beginning how late in the project you can make changes. In other words, if the toolmaker is cutting metal and you make a dimensional change, you most certainly will pay dearly for it.

IV. CHANNEL OF COMMUNICATION

Next, be sure your company has only one channel of communication to the moldmaker and that that channel includes you. Nothing is more embarrassing than to find out your engineer has called the moldmaker and made extensive changes that disrupt the time schedule, increase costs, and generally upset the whole project.

Some situations have evolved so badly that when the tool came into the molder to be sampled, he did not even recognize the part. This can lead to much ill will, a lot of lost time, and big bucks.

Some molders, if they know what they're doing, will insist their toolmakers make absolutely no changes without their approval. This method helps to protect everybody and is highly recommended.

V. FIRST-TIME MOLD TRYOUT

One of the most emotional and sometimes difficult times any structural foam customer can have is during the sampling of a new mold. After the initial purchase order for tooling is placed, there is a long waiting time while the mold is being built. During this time when the customer has committed substantial monies for down payment, when the designer or engineer is chewing on his nails, waiting for his brainchild to be born, when precious days of waiting are going by, it is understandable that everyone involved is anxious. During all this one must keep in mind that you are usually paying for and will receive a unique, one-time-only piece of machinery, that will need fine tuning before it will run hundreds of acceptable parts.

On the first tryout, it is sometimes best to let the moldmaker and molder handle this themselves. This is often hard to do since you have waited so long and you want to be there for the great moment. Be sure you give the molder some room. There is nothing that activates "Murphy's Law" more quickly and thoroughly than the tryout of a new mold with the customer and others breathing down the molder's neck. Keep in mind that the average number of samplings on a structural foam tool before it is approved is three times. It could be many more.

Tools are often built with too-small runners and gates so that they can be opened up as needed. Each part is usually different and no one knows the right system before the part is first molded. Oftentimes during the first sampling, the parts are not even filled. Short shots, stuck cylinders, and hundreds of other things can make these first shots look like nothing you would recognize, certainly not the new part on which you have spent so much time and money.

The moral is: If you have to be on hand for the first tryout, be kind, be aware of the problems, and let the molder and the toolmaker work them out. Your chances of getting an acceptable first article in the shortest period of time depends on it. Better still, keep your people away till the molder has things under control and is ready to show off his capabilities . . . and your new mold.

VI. FIRST ARTICLE INSPECTION

Once you have received the first full shots from your new mold, you will want to take them back to the plant for a complete inspection. Some customers require their molders to provide a first article inspection report. However, this is usually only a backup report to their own.

Molders will always check major important dimensions first off, to confirm shrinkage. When reviewing a set of parts, it is certainly most important that they fit together. Often, the actual dimension is not critical, only that they fit one to the other.

In any structural foam project, the molder usually contracts to make parts to print. Some customers feel that this means every single dimension on a print will be to tolerance or they will not accept the part. Others take a more tolerant view—that only the dimensions that are critical need to be to tolerance and that others that are noncritical can be changed on the drawing to bring the part into tolerance.

This last approach seems to make the most sense since it is often an impossible feat for all dimensions to be to print the first time the mold is sampled. You can certainly hold the molder to the letter of his contract and force him to bring *all* dimensions up to the existing print. He will no doubt do this, but it will cost you time and often keep both the toolmaker and molder from getting the part into production in the shortest period of time.

What is really needed here is the team effort, where all parties are working to produce a workable and reproduceable part in the shortest and least expensive way.

VII. QUALITY CONTROL

The intense desire for both the molder and the customer to deal only with acceptable parts creates some strange procedures. Under the guise of quality control or quality assurance, elaborate, expensive, nonproductive personnel groups are set up by both parties, often with little success.

Quality control actually starts with good communications and a thorough understanding of what the part does, looks like, and how it is produced. Quality control with a structural foam molder usually starts with the machine operator. If this key person understands and takes an interest in what is coming off the machine, you can go a long way in obtaining acceptable parts.

A. Perfect Parts

The key word here should probably be *acceptable*. Many OEM, especially in the electronics industry, look for the "perfect part." This is no doubt due to the need to inspect and be sure of the quality of each small electronic component before it becomes a part of a built-up system. It is no wonder that this carries through to the housing. Usually the same quality control group under the same quality control guidelines is responsible for both. Unfortunately the problem of supplying a continuous stream of perfect parts is very close to impossible. Paint textures vary from the paint supplier and defects in molding can get through.

B. Communication

The way to keep these problems under control is again communication. Signed off samples of acceptable parts, both finished and unfinished, should be available to both the molder and the OEM. Color chips that show acceptable color and also acceptable variation in texture should also be available at both locations and the paint facility.

This also brings up the need for one channel of communication between supplier and customer. A situation where more than one person in a department has the capability to make changes is a disaster waiting to happen. Usually the channel is purchasing to sales. However, whatever it is, be sure it is defined early in a project and then enforce it.

Fortunately, in either structural foam or injection molding, once the mold is producing parts to print, the reproduceability is excellent. Areas that can change are holes missing due to broken pins, inserts moving causing minor dimensional changes, and short shots causing short bosses. By reviewing closely the areas on a part that can be problems, you should be able to spot and eliminate early these trouble spots.

Most molders want only to ship good parts. Unacceptable parts on a customer's dock are expensive to all concerned. An offer to make these parts acceptable by providing labor or whatever, will go a long way toward cementing a molder-supplier relationship, especially if it is at the beginning of a project where both parties are on a learning curve.

C. Learning Curve

Another thing to remember is the existance of a learning curve on any project. Be patient and let the molder work out the early problems. Once acceptable production patterns have been set, you should experience very few of the early problems.

Finally do not require tighter control than you actually need. The goal for both companies should be the on-time delivery of acceptable parts with the least amount of cost. Too tight a quality control program will raise the cost and increase the problems.

The technique of using quality control to send back parts you ordered but do not need can also be expensive. Vendors normally catch on to this ploy early and you may lose a good source. Remember, if you are getting acceptable parts from a molder and lose him, your next source may not be as good.

VIII. MOVING MOLDS FROM VENDOR TO VENDOR

This situation is one dreaded by both purchasing personnel and molders. It fortunately does not happen often, and if you have done your homework it should not create the animosity and problems that it often does. Once the decision has been made to more tooling, one should forget the reasons for moving it and handle the situation like any business arrangement.

A. Financial Situation

First, review the financial side. Most molders will insist that you settle all outstanding bills before moving tooling. In some states this is not allowed, however, this is changing. Clear up the bills and you will normally find little resistance to movement of the tool.

Table 7.1 Mold Progress Report

Date: ____________

Customer: ______________________________ Job No.: ____________

P.O. No.: __________ DWG. No. __________ Rev.: ____________

Description: ____

Date of order: ________________ Acknowledged delivery: ________

Mold description : ____

Started | Proposed Completion | Revised Completion | Completed

Date of review																
Mold design																
Duplicating models																
Mold materials																
Cavity machining																
Core machining																
Water lines																
Inserts																
Slides/pulls																
Polishing/fitting																
Ejector assembly																
Final assembly																
Inspection/ship																
Overall mold completion																

Remarks ____

____

Prepared by: ______________________________

B. Mold Examination

The next important step to take is to send a knowledgeable person to examine the molds before they leave the molder's shop. Contact the molder early enough ahead so that he can have the molds open for your inspection. Ask him to also show you the last shots made from the tooling (most molders keep the last shot). By reviewing both part and mold, you can determine such things as coined parting lines, surface condition of cavity and core area, insert condition, if applicable, and general condition.

If a problem is found, it is not out of line to ask the molder to have the problem taken care of. Once you have reached an agreement on this, you can then make arrangements to move the tool. Most tooling can be shipped on pallets. However, you may wish to have it crated if it is a complex tool with cylinders protruding. Do not forget to take out adequate insurance on the tool during shipment.

C. New Vendor Problems

You can ignore all of the above and have the molder simply make the move himself. However, do not be surprised if your new molder's first comments are, "We didn't expect the mold to be in this condition. We'll have to review our quote."

Molders who quote on existing tooling will rightfully add this statement onto their quotations. It is only fair that the new molder get a chance to examine the tool before he is tied to a final price on the part. Tooling or the condition of same can have a great effect on the cycle, the amount of part clean-up, and other parameters. Be prepared to pay for some mold rework and clean-up when a mold is moved. Be sure your molder is allowed to do (within reason) what he wants to the tool, but then hold him responsible for the part price quoted.

8

Tooling of Structural Foam *

I. INTRODUCTION

One of the most expensive and misunderstood parts of the structural foam product design function is the development of tooling. This chapter, prepared by Mr. Dan Swistack, Sales Manager for Hoover Universal, Inc., East Longmeadow, Mass., a large manufacturer of all types of plastic tooling, defines this situation.

There is no substitute for a close relationship between moldmaker, molder, and OEM. The problems created by poor communications between the three main participants in a foam project are too numerous to mention.

II. MOLDS FOR FOAM PROCESSING

Molds for foam processing have run the gamut in the past 10 or so years due to increasingly complex part geometry, tighter tolerances, and higher demands for quality in the molded parts. In the formative years of foam molding, the major tooling medium was cast aluminum and kirksite, whereas in the current era, the molds are approaching injection-mold standards.

A. Value of End-User Involvement

More than ever before, it is worth the while of end users to become actively involved in the tooling purchase. This direct

*Source: Ref. 1.

FIGURE 8.1 A toolmaker works on one of the large machined aluminum molds which are common in the structural foam industry. Several mold-making companies in the United states have specialized in the production of these giant tools.

involvement with the molder and moldmaker will help to produce a mold that will sufficiently handle the job at hand and quite possibly save time, money, and problems during the process. It is therefore hoped that the following information will provide the end user, molder, designer, and others with enough knowledge to help them during the mold-buying stage of their project.

B. Mold Quality

It is not only practical, but essential to try to get the best possible mold for the money, but it makes no sense to buy an injection quality mold for a product that will require 5–10,000 parts and then be made obsolete by a newer design. Therefore, it is important that the triumvirate (end user, molder, moldmaker) understands what is required or expected of the product to give the desired results at the lowest possible cost. The key individual in this case is the end user. He has to provide the molder and moldmaker with a part drawing or concept of the product he wishes to have molded. (See Fig. 8.1.)

C. Information Needed

Along with this drawing, the following information should also be provided so that the job can be quoted accurately by the molder and moldmaker.

1. Number of parts per year
2. Tolerances required in the finished part
3. Surface finish (i.e., as molded, painted, textured, etc.)
4. Areas where gates are allowable
5. Surfaces that will have to match other parts
6. Plastic material to be used
7. Areas where ejector pin marks are acceptable
8. Allowable draft angles

Once the molder and moldmaker have this information, then they can determine what type of mold will best fit the application at hand.

III. MOLD MATERIALS

In general, one of the first considerations in the making of a mold will be the material used to construct the tool. Selection of a mold material depends upon the expected life of the project, (i.e., number of parts per year and number of years the mold will be in service); the molding process (i.e., high-pressure foam, low-pressure foam, reaction injection molding, thin-wall foam, etc.); complexity of product design and other special considerations (i.e., cooling, ejection, shut-offs, etc.).

A. Advantages and Disadvantages

The following is a listing of the types of metals generally used for mold construction.

TABLE 8.1 Mold Construction Materials

Mold Material	Advantages/Disadvantages
Metal-filled epoxy	Not recommended for tight tolerance parts Primarily for prototype molds; 1–10 parts (although more parts are possible) Hard to control parting line flash Poor conductor of heat Inexpensive to make Molds can be made in a matter of weeks Easily damaged due to softness of material Surface finish of molded parts is poor
Spray metal with epoxy backing	Primarily used for prototype molds (up to 100 parts) or low-production reaction injection molding (RIM) molds (up to 5000 parts) Hard to control parting line flash Poor conductor of heat Inexpensive to make unless the part geometry is complex; at this point cast aluminum or kirksite may be necessary Molds can be made in a matter of weeks Easily damaged due to softness of material Long cycle times Surface finish of molded parts is poor Not recommended for tight-tolerance parts
Kirksite	Used for prototyping and low production in all types of foam processing Molds can be made in about 8 weeks

TABLE 8.1 (Continued)

Mold Material	Advantages/Disadvantages
	Soft material (therefore, either the mold will require a lot of maintenance or the molded part will require abnormal secondary finishing) Generally inexpensive to make Difficult to repair or alter this type of mold Poor conductor of heat Not recommended for tight-tolerance parts
Cast aluminum (grades commonly used are A356 and 319) (see Fig. 8.2)	Excellent reproducibility of pattern details Used for prototype molds and low-production molds in all types of foam processing Porosity in the metal is common (can cause blemishes on molded parts) Molds can be made in about 8 weeks Soft material (therefore, either the mold will require a lot of maintenance or the molded part will require abnormal secondary finishing) Generally inexpensive to make Good conductor of heat Easy to repair or alter this type of mold Not recommended for tight-tolerance parts
Machined aluminum (grades commonly used are 6061-T651; 6061-F; 7075-F; 7075-T651 and 7000 series)	Used for prototype, low-production, medium-production, and, in some cases, high-production molds for all types of foam processing Excellent conductor of heat Easily machined Costs approximately 15–30% less than steel in most cases Easy to repair or alter this type of mold Harder than cast aluminum or kirksite, but generally, less than 50% as hard as steel

TABLE 8.1 (Continued)

Mold Material	Advantages/Disadvantages
Machined steel (grades commonly used are P-20; 4140 and 4130)	Used for all types of production molds Good conductor of heat Generally more expensive than other mold materials Difficult to machine Difficult to repair or alter this type of mold Excellent mold material Capable of absorbing a great deal of abuse

Beryllium copper (BeCu) has been used to a small extent in the production of molds for the furniture industry. The reason for its popularity in this industry is that cast BeCu has the highest degree of pattern reproducibility of any metal and is used to produce foam parts that are designed to replicate wood products, hence the need for molded-in wood grain. However, the price of BeCu has skyrocketed and is rarely used today.

IV. GATING

A. Thermoplastics

Positioning of the point of plastic material entry is our next demanding task, no matter what the molding process. In thermoplastic foam processing, the plastic material is fed into the part through one or more nozzles, depending on whether the machine is a single or multinozzle machine. The common practice is to place the nozzle seats in the tool so as to gate directly on the part in an area that is not visible in final assembly. It is beneficial to gate directly over or into a rib in the part. This method increases the flow of the material as the rib acts as a natural runner for the plastic material. If it is impractical to gate directly onto the part, tab or fan gates are used. The tab gate is usually a half-round approximately 1/2-in.-deep × 1-in.-wide with a 1/4-in.-deep × 1-in.-wide entry into the part. The fan gate is similar to the tab gate except that the width of the entry into the part will vary depending on the part configuration. In all cases where gates or runners are used to get

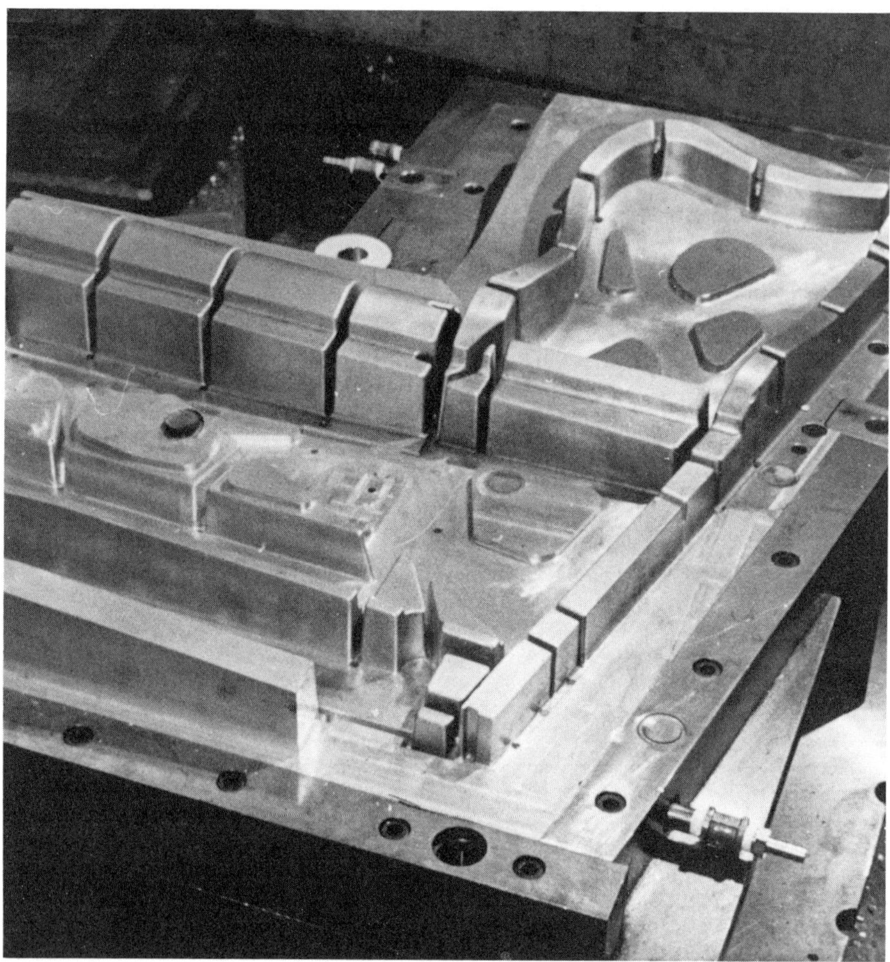

FIGURE 8.2 The cast aluminum core for a structural foam tool which is designed to produce the inner frame of an upholstered chair.

the material into the cavity, a postmold operation is required to trim off these gates and runners.

B. Reaction Injection Molding

In reaction injection molding (RIM), the plastic material is fed into the mold through a high-pressure impingement mix head and there is generally only one gate required to get the plastic material from the mix head to the cavity. There are a number of aftermixer and gate designs used to obtain the proper flow of the material into the cavity (see Fig. 1). In all cases, the gate should be positioned at the lowest point in the mold to allow the material to push the entrapped air to the highest point for venting and to keep turbulence of the material at a minimum. This will ensure that the parts are molded with as few surface imperfections as possible.

V. VENTING

The location and quantity of vents in any mold is important to the production of a good part. The quicker entrapped air is evacuated, the longer the possible flow length of the plastic material and the better the part finish. Nozzle and gate locations, fill rates, and melt paths will determine the location and number of vents required.

VI. COOLING

Cooling line placement is also important to the production of a quality tool for all foam processes. Although it is true that foamed material will give up heat at a given rate, it is desirable not to have any hot spots, which could cause surface imperfections in the product, as well as trying to keep the cycle time as short as possible. If the skin strength is not sufficient to withstand the internal pressure of the foaming gas, an expanding bubble will form on the part surface that could ruin its appearance or interfere in the assembly of the product. For this reason good temperature regulation is a *must*.

A. Cooling Line Dimensions

Cooling lines that are drilled into machined metals are generally 9/16 in. in diameter and between 1 1/2 to 2 1/2 in. apart. For best results, the lines should be as close to the plastic material as possible and these lines should be connected to flow in numerous short circuits to allow for maximum regulation of the overall water temperature. Baffle blades, water cascades (bubblers), and heat pins are used in tight places where deep cores or core pins are involved. These allow for a continuous flow of water where lines cannot be cross-drilled to create a circuit. In the case of epoxy molds or cast metal molds, 1/2-in. O.D. stainless steel or copper tubing is used and it is generally cast in place.

VII. EJECTION

Part ejection is very similar to conventional injection-molding tools. The standard types of ejection utilize pins, bars, or plates that can be activated by normal operation of the molding machine or by use of hydraulic cylinders, chains, or pull rods; the latter two being least desirable due to a lack of accuracy that could cause the ejector plate to bind. For the most part, the ejector system consists of a set of steel support rails, which are high enough to allow for an ejection stroke sufficient to push the part completely off the core, a steel clamp plate, steel ejector plate, and steel ejector retainer plate. All the major components of the ejection system are standard items available from any of the mold component suppliers such as DME and National. For the majority of foam molds, four 1 1/4-in.-diameter ejector plate guide pins and four 1-in.-diameter ejection return pins are sufficient to provide a smooth-working, trouble-free ejection system. It is extremely important for the moldmaker to know on what type of molding machine the tool will be used, since the knockout arrangement differs with each machine. To run the press on an automatic cycle, it is essential that the ejection stroke of the mold be sufficient to strip the molded part entirely from the core, letting it fall freely to the part removal area below.

VIII. CORE PULLS AND SLIDE MECHANISMS

One advantage of utilizing plastic in design is the ability to eliminate numerous pieces by incorporating as many features as possible into the molded part. In some cases, this may involve little more than a core pin to create a hole in the product and can be accomplished in the normal motion of the molding machine. However, in the case of features that are contrary to the normal parting line, a core pull or slide mechanism is needed to allow the part to be removed from the mold. It may be simply a core pin attached to a cylinder to create a hole in a side wall of the product or it could necessitate moving an entire wall of the mold. Whatever the case, care should be taken to provide for safe, trouble-free operation of every core pull or slide mechanism.

A. Methods of Actuating

Common methods of actuating these mechanisms are hydraulic or air-operated cylinders, cams, angle pins, lifters, and springs. In all cases, each moving piece that is installed in a mold should have some means of position lock or limit switch to position the slide in its proper molding condition and to prevent damage to the mold or operator. The sliding pieces and the surfaces on which they slide should be constructed of dissimilar metals, usually steel and bronze. There should also be some means of lubricating these slides to keep them working smoothly. Electrical safety interlocks are a must for all slide actions.

IX. SUMMARY

It is hoped that the preceding information will help you in finding a source capable of producing your molds with the quality you expect. Price does not equate directly with quality. There are a number of items that dictate price such as the work load of the tool shop, the type of machine tools in the shop, and the expertise of its moldmakers. Therefore, it is to your benefit to be involved in the selection of a tool source, because, after all, it's you, not the molder, who is buying the mold.

REFERENCE

1. Daniel J. Swistack, Hoover Universal, East Longmeadow, Massachusetts 01528; an unpublished paper.

9

Designing with Structural Foam

I. INTRODUCTION

Various parts of the design picture have been presented in other chapters. This is an attempt to bring them together in a step-by-step procedure for developing a structural foam application.

II. REVIEWING FUNCTION

The main objective for the designer of any specific product is to design each and every part so that the sum of the parts provides a completed, workable assembly at the least possible cost. To accomplish this, one must first evaluate the function of the part and then define the parameters within which it must function. These parameters include use, temperature, number of cycles, load, general environment, aesthetics, and others.

III. PART DESIGN

Once this criteria has been established, one can then set out to design the part or parts to accomplish the function. One first designs the part to perform the function, then modifies it according to how it is to be produced and from what material.

IV. MATERIAL AND PROCESS SELECTION

Once the preliminary design of a part or parts have been developed and put on paper, you can then make the material selection based on use requirements. This done, you have narrowed the processing selection down to a relatively few choices. From here, the economics will dictate your choice. If the need is for only a few parts, the choice would probably be to machine then from bar or rod stock. Higher volumes require tooling with the initial high cost, but less expensive piece prices. In order of tooling costs, the more available techniques would be:

Vacuum forming
Structural foam
Sheet molding
Injection molding

Other factors, such as part size, will also influence the choice.

V. CONFORMING TO THE PROCESS

Once material and process have been selected, the designer must then modify his designs to conform with the process. Such items as sharp corners, draft, and gate locations are all factors to be considered. Assembly of the parts must also be taken into consideration at this point.

VI. DESIGN SUGGESTIONS

A few suggestions for designing for structural foam applications are given below. These ideas come from the new structural foam booklet put out by the Structural Foam Division of the Society of Plastics Industry [1] and the design manual published by General Electric Co. [2].

A. General Design Rules

Designing with foam is similar to designing for any molded product. Some general rules include:

1. Keep undercuts to a minimum unless you are prepared to pay for slides, cams, and cylinders.

2. Use as generous a draft as possible on vertical walls while still keeping wall thicknesses as uniform and as thin as possible.
3. Keep wall sections uniform where possible and go from thick sections to thin with material flow, (1/4 in. wall thickness is optimum for most thermoplastic foam molding processes). Thermoset material usually requires greater than 1/4 in., but can be molded at lower densities.
4. Keep material flow length to a minimum.
5. Molded foam parts are still notch sensitive, so use generous radii whenever possible.
6. Coring thick sections is a winner in two ways. First, the reduction of material; second, this same coring also cuts cycle time and reduces machine cost.

VII. MORE ON CORING

Most molders charge for material plus an hourly rate. Coring helps reduce both. One should keep in mind that cycle times on structural foam are longer than on solid materials due to the expansion tendency and the fact that the foamed materials are self-insulating. This cycle time is determined by the thickest sections in any part. This can prove very expensive in parts having thick sections that are unneeded.

Another factor to be considered with coring is that vertical skins formed by the core walls increase the stiffness of the cross section. This is the old beam design at work and is a plus to be aware of in the foam design.

VIII. TOLERANCES

As with any molded item, tolerances and the ability to hold them has always been in question. The old adage of "don't demand tight tolerances unless you need them," is definitely true with structural foam. When close tolerances are needed, most of the engineering materials will provide good reproducibility. This, of course, assumes the mold is right in the first place. Tolerance specification is extremely important in product design as it can directly affect both part cost and the ability to perform properly. The practice of making blanket tolerances on a drawing can result in unnecessary cost both in the part and the tool. (See App., Sec. VIII.)

The final dimensions on a part are affected by:

Thermal expansion and contraction
Processing conditions
Mold dimensions and design
Part configuration
Material selection

IX. UNIFORM WALL THICKNESS

Maintain uniform wall thickness throughout the part. Thick sections will not only cause increased cycle times, but may contribute to part warpage. Minimize restrictions to material flow. Omit ribs unless absolutely necessary and orient louvers in the direction of the flow. Above all, remain within the flow lengths of a material.

X. FILLERS

The addition of fillers such as glass or talc will reduce mold shrinkage and this technique can often be used to bring tolerances into specification. Again, experience counts for a great deal, and most experienced structural foam toolmakers and molders have developed a pretty good feel for adjusting shrinkages. In most cases, the end user only stipulates the part dimensions required and leaves the shrink juggling to the moldmaker and molder.

XI. COEFFICIENT OF EXPANSION

Contraction and expansion of plastic parts should also be considered in any design situation where plastic is secured to metal. Plastic materials will expand much more over a given temperature gradient than a metal part and, if tied together by fasteners, will bow the system out of shape. Slotted fastening holes will normally take care of this problem and should be considered where appropriate. Special design problems are encountered with the molding of larger and larger parts. Shrinkage over a 4–6-ft part can be excessive and quick ejection from the mold is critical to keep the part from tearing itself apart. (See App., Sec. I.)

XII. RIBS

Ribs can be used to increase the rigidity and the load-bearing capability of a structural foam part without increasing the wall thickness and part cycle time. Since structural foam parts are much less subject to sink marks, thicker ribs can be incorporated.

A. Guidelines for Rib Thickness

A general guideline for rib thicknesses as a function of the surrounding wall is as follows:

Wall Thickness (in.)	Rib Base Thickness as % of Surrounding Wall
0.157–0.175	75
0.176–0.215	85
0.216–0.300	100
0.300	120

B. Draft Angles

Draft angles on the rib should be about 0.5–1.5° while fillet radii at the base should be 0.030 R to 0.050 R. Ribs should be added only when necessary because when oriented improperly, they sometimes contribute to part warpage.

XIII. SINK MARKS

Sink marks, so prevalent in injection molding, are less a problem with structural foam. However, large flat surfaces backed with ribs can exhibit minor sink marks opposite ribs. Reducing the rib thickness to something less than wall thickness will help to eliminate this problem.

XIV. WALL THICKNESS

The structural foam process permits molded parts with sections thicker than can be realized in injection molding without sink marks and warpage problems. Traditionally, structural foam parts were designed with 0.250-in. (6.35-mm) wall thicknesses. Now with modified engineering resins, parts can be designed with wall sections as low as 0.157-in. (4 mm). The design criteria of a structural foam part must be considered before choosing the optimum wall thickness and material for an application. (See App., Sec. III.)

XV. DRAFT ANGLES ON PARTS

As in injection molding, draft angles are necessary in structural foam molding. Because of the lower pressures involved in foam molding, smaller draft angles can be tolerated in certain cases. Generally, an angle of 0.5–3° will provide sufficient draft to release a part. Textured surfaces on sidewalls generally require an additional 1° draft per 0.001 in. depth of texture. For best results, consult your engraver for depth versus pattern before specifying the draft requirements.

XVI. FILLETS AND RADII

Sharp corners create points of stress concentration and restrict material flow in a structural foam part. They are often a major cause of part failure. Use as large a radius as possible on inside and outside corners to minimize this stress concentration and aid in mold filling. In most parts, the minimum inside radius should be 0.060 in. If the section is under load or subject to impact, a minimum radius of 0.125 in. should be adopted. A radius equal to 0.6 times the wall thickness will provide a desirable fillet for practical purposes.

XVII. BOSSES

Bosses can be easily incorporated into structural foam parts to accept fasteners and support components. In many applications, the addition of molded-in bosses, mounting pads, stand-offs, and retainers can replace costly brackets and miscellaneous small metal part assemblies. In general, boss diameters should be 2.0 times that of the cored hole. This recommendation will vary somewhat, depending on resin used and boss wall thickness. Bosses should be cored whenever possible to prevent the formation of a thick section in the part. Generous fillet radii should be used at the base of the boss to avoid stress concentration and resist torque loading.

XVIII. TRANSITION SECTIONS

Transition sections from thick to thin walls are more easily achievable without sink marks in structural foam than in injection molding. Still, uniform wall thickness should be maintained whenever possible to minimize restrictions to material flow.

Transition from thick to thin walls should be tapered for proper processing of the structural foam part. In molding parts with wall sections of varying thicknesses, it is often better to gate the part in the thin section and allow the material to flow into the thicker area.

XIX. LOUVERS

Since thermoplastic structural foams are insulative in nature, oftentimes louvers must be designed into parts to get the heat generated by enclosed components to the outside. Louvers often create restrictions to flow and care must be taken in their design. Louvers should be oriented in the direction of flow. If not, a runner should be provided down the middle to allow material to flow to the outside.

XX. HINGES

Properly designed, integral structural foam hinges offer fatigue strength comparable to metal, while eliminating costly bracketry and assembly time. Hinges can be designed to be either hidden or visible, depending on their location in the part.

XXI. SNAP-FITS

The superior rigidity and strength of structural foam parts permits increased utilization of snap-fits for assembly and for mounting heavy components in bases. A quick and extremely economical assembly method, snap-fitting eliminates the need for added screws, brackets, and fasteners, significantly reducing labor and assembly costs.

XXII. DESIGN INFORMATION FROM OTHER SOURCES

Sections in the Appendix are taken from an excellent design manual by Borg Warner Chemicals and are typical of the information available from material suppliers. While this information was prepared for Borg Warners Cycolac® ABS, it applies to many other materials used in structural foams. The selections selected were particularly apropo to structural foam.

Other sources of information for structural foam design ideas include:

SPI Structural Foam Division Annual Conference
Trade journals such as *Plastics Design & Forum*
Industrial designers
Existing applications of structural foam

XXIII. SUMMARY

In summary, one must find design information where one can and, above all, never take what is found for granted. Test information is always subject to interpretation, and unfortunately it is usually only the positive that finds its way into our publications. Good plastics design, whether it is structural foam or another plastics form, is built on common sense. Consult as many sources as is practical, list your options, and make the best decision from what you have found.

The following disclaimer is added at the request of the General Electric Company and applies to all design information supplied by them:

Inasmuch as the General Electric Company has no control over the use to which others may put the material, it does not guarantee that the same results as those described herein will be obtained. Each user of the material and the compositions described herein should make his own tests to determine the material's suitability for his own particular use. Statements concerning possible or suggested uses of the material described herein are not to be construed as constituting a license under any patent covering such use or as recommendations for use of the infringement of any patent.

REFERENCES

1. Structural Foam, S.P.I. Structural Foam Division, 4th printing, 1984.

2. Engineering Structural Foam Guide, General Electric Co., Plastics Operation, One Plastics Ave., Pittsfield, Massachusetts, 01201.

Appendix: Design Tips

I. DESIGN TIP NUMBER ONE—THERMAL EXPANSION

A. Attaching Cycolac® ABS Plastics to Materials Having Different Thermal Expansion Coefficients

The linear thermal expansion coefficient of a material describes the material's dimensional change to a fluctuation in temperature. Typically, the values are listed in data sheets as the change in inches per inch of length per degree changes in temperature over the range of −22°F to 86°F. For the various Cycolac ABS grades, values range from a low of 3.7×10^{-5} to 7.2×10^{-5} in./in./°F.

To determine the change in length of a part, it is necessary only to multiply the part length by the change in temperature and the thermal expansion coefficient of the material used. If the part is free to expand and contract, then its thermal expansion property is usually of little significance. However, if it is attached to another material having a lower coefficient of expansion, then movement of the part will be restricted. A change in temperature will then result in the development of thermal stresses in the part. The magnitude of the stresses will depend on the temperature change, the method of attachment, and the relative expansion and modulus characteristics of the two materials at the temperature of interest.

For purposes of reference, the thermal expansion coefficients and modulus properties for the materials most commonly used with Cycolac® ABS plastics are listed as follows:

Material	Coefficient of Expansion	73°F Modulus
Wood (pine)	0.3×10^{-5} in./in./°F	12.4×10^5
Steel	0.6×10^{-5} in./in./°F	300.0×10^5
Aluminum	1.3×10^{-5} in./in./°F	100.0×10^5
Cycolac LS	6.3×10^{-5} in./in./°F	2.5×10^5

Note: The plastic material modulus at the temperature extreme is used, *not* the 73°F modulus.

II. DESIGN TIP NUMBER TWO—PRODUCT COLOR—HOW IMPORTANT IS IT?

Color selection for a thermoplastic product may be important for reasons other than aesthetics. It is a well-known fact that the surface temperature of an object exposed to sunlight is dependent on its color. If the product is for exterior use, then the selection of color may be critical from this standpoint. This is especially true in the case of products such as golf cart bodies or boats. The everyday use of these types of products will involve considerable contact with a person's body. For these types of products, it is essential that the surface temperature build-up be held to a level that "feels comfortable to the touch."

To aid the designer in making this choice for products made of Cycolac® ABS, the list is provided below which shows typical surface temperatures measured on various colored ABS panels. These measurements were made on panels exposed to an actual outdoor environment at an ambient air temperature of 100°F. Air flow around the panels was not restricted in any way.

Panel Color	Surface Temperature
White	120°F
Light blue	127°F
Medium blue	137°F
Dark blue	144°F
Medium red	128°F
Dark red	137°F
Black	148°F

The estimation of the influence of surface temperature build-up should include consideration of the total product assembly. In applications where air flow around the part is restricted, surface temperatures will be somewhat greater than those listed above. Also, some applications in order to obtain added stiffness require that the Cycolac® ABS be backed by urethane or polystyrene foam. In these cases, the excellent insulating characteristics of the foam materials result in even greater surface temperatures, as is illustrated below.

	Surface Temperature	
Panel Color	Unbacked	Foam Backed
White	120°F	127°F
Light blue	127°F	137°F
Medium blue	137°F	157°F
Dark blue	144°F	174°F
Medium red	128°F	142°F
Dark red	137°F	169°F
Black	148°F	173°F

Ambient air temperature = 100°F

Another area in which high surface temperature can play an important role is part warpage. All materials expand when subjected to an increase in temperature. The property that describes this characteristic is called the materials coefficient of linear thermal expansion and is expressed as the materials change in length per unit length per degree change in temperature.

Consider the example of two Cycolac® ABS parts, one black and one white, which are identical with the exception of color. Based on the previously listed surface temperatures, exposure of the parts to a sunny outdoor environment will result in greater expansion of the black part due to a greater surface temperature build-up. If expansion of the part is restricted in any manner, such as being attached to a framework of a material having a lower thermal coefficient, then stresses will develop. Warpage, or distortion, is one of the ways in which the part will tend to relieve these stresses.

In summary then, color, which directly influences surface temperature, has an influence on part expansion and potential part warpage. The data presented here show these potential problem areas may be minimized through the use of light pastel colors.

III. DESIGN TIP NUMBER THREE—EQUIVALENT RIGIDITY OR STIFFNESS OF MATERIAL

In determining the stiffness of a product made from a specific material, it is necessary to know the flexural modulus characteristics of the material. For plastics, these data should be taken from the manufacturers' published, "Typical Property Data Sheets." For other materials (e.g., aluminum, steel, FRP, wood, etc.), these data can be found in an engineering handbook. Simply examining modulus values for the materials can be somewhat deceiving. For instance, if we compare Cycolac® ABS, grade LS, to aluminum, we have flexural modulus values at 73°F (23°C) of 250,000 psi, versus 10,000,000 psi, respectively. At first glance from a plastics standpoint, this looks impossible. Does this mean that to equal the rigidity of aluminum, the plastic must be 40 times thicker? Fortunately for plastics, this is *not* what it means.

Calculating equivalent rigidity in bending of simple beams of different materials can also be accomplished by utilizing the following equation:

$$t_1 = 3 \frac{E_2 (t_2) 3}{E_1}$$

Where E_1 = Flexural modulus of "new" material (at the temperature and/or time of interest).
E_2 = Flexural modulus of "old" material.
t_1 = Unknown thickness of "new" material.
t_2 = Thickness of "old" material.

IV. DESIGN TIP NUMBER FOUR—CHEMICAL RESISTANCE CONSIDERATIONS

Cycolac® ABS has performed successfully in a great many application areas because of its good chemical resistance properties. Success was attributable in most instances to the thoroughness with which end-use environments were identified and evaluated for the candidate Cycolac® ABS grade.

To minimize the risk potential for new applications, all media to be in contact with the ABS grade planned for use in a proposed application—media such as lubricants, gaskets, upholstery materials, foodstuffs, cleaners, polishes, paints, cosmetics, etc.—should be evaluated for compatibility under anticipated end-use conditions.

Experience has taught that the term "chemical resistance" means different things to different people. The chemical resistance of Cycolac® ABS grades is measured by Borg-Warner Chemicals by three different types of tests. All of these tests are influenced by temperature, media concentration, duration of exposure, and type of loading.

ABS plastics actually represent a broad variety of materials (including high-heat materials, flame-retardant grades, expandable grades, alloys with polycarbonate and PVC, besides a variety of general purpose grades). Consequently, chemical resistance performance can vary between grades as well as between ABS products from different manufacturers.

A. Environmental Stress Cracking

When thermoplastics are exposed to different chemical media while under load (or stress), crazing or embrittlement may result. This phenomenon is tested using various techniques, all involving the simultaneous exposure of the plastic to the chemical and a stress. Under these conditions, the plastic may be characterized as exhibiting a "critical stress" below which the chemical media has no apparent effect. If this critical stress level is not exceeded while the plastic is in contact with the chemical, the plastic may be used for structural applications. This assumes proper consideration has been given to other application end-use performance criteria such as loading rate, temperature, etc.

B. Simple Immersion (ASTM D543)

Swelling or Solvation

With simple immersion (generally in organic liquids), there may be a weight change due to absorption of the chemical. This is often accompanied by a softened or tacky surface and changes in physical properties, but not an actual chemical reaction. The plastic is under no load during the test.

Etching or Chemical Attack

With simple immersion (generally in concentrated inorganic liquids), there may be actual chemical attack and decomposition of the plastic. This is accompanied by a weight change and changes in physical properties. The plastic is under no load during the test.

C. Staining (ASTM D2299)

With staining, there may be surface discoloration or a change in gloss with little or no change in physical properties. The color of a plastic can affect the noticeability of stains. The plastic is under no load during the test.

It is important to note that there can be an interchange of ratings for an ABS grade or a chemical in going from one test to another. Since each test is measuring a different set of performance criteria, this is understandable—for example, a stress-cracking agent for a particular ABS grade may not be a staining agent for that grade.

To project accurately the performance of a material in an application, it is important to review the data from each chemical resistance test category and reflect upon application end-use requirements that have been defined. Ultimately, the manufacturer must assure himself of the adequacy of his chemical resistance decision(s) for an application through simulated-use tests, or controlled field studies.

V. DESIGN TIP NUMBER FIVE—DESIGNING TO ACCOMMODATE TOXICITY REGULATIONS AND STANDARDS

The need for information by designers and producers of products which must meet certain regulatory agency approval is evident. With this need in mind, this design tip is being published in an effort to assist our customers with this problem.

The usual toxicity regulations or standards that must be considered in designing an end-product, and in choosing the proper ABS materials for that product, are from these sources:

Government

1. *Food and Drug Administration (FDA)*—principally regulation 21 CFR 181.32, formerly 121.2010, for repeated-use food-contact applications.
2. *U.S. Department of Agriculture*—Meat and Poultry Inspection Program—general preapprovals are issued by USDA for various ABS grades, depending on colorants and on type of applications.
3. *Consumer Products Safety Commission*—Federal Hazardous Substances Act, Regulation 16 CFR 1500.

Nongovernment

4. *U.S. Pharmacopeia* Section XX (Medical Devices)
5. *National Sanitation Foundation (NSF)*

Material acceptability for use in contact with food, potable water, and certain kinds of uses in toys and medical devices cannot be divorced from the design and the construction of the device. It is necessary that all applicable regulatory standards be identified early in planning, and the proper ABS grade chosen to meet both nontoxicity and performance needs.

In Conclusion

Key points to bear in mind when considering any plastic material for use where performance to standards is necessary:

Know the specific conditions of the anticipated use. Keep up-to-date on current regulations that might affect the marketability of the product. Processors should be aware of the proper storage and use of regrind; and understand their responsibility for material composition whenever in-house coloring or compounding is done.

VI. DESIGN TIP NUMBER SIX—THE INFLUENCE OF UNDERWRITERS LABORATORIES STANDARDS ON PRODUCT DESIGN AND MATERIAL SELECTION

The Underwriters Laboratories, through their "Standards for Safety," exert major influences on the design of appliances and the selection of plastic materials for enclosures. It is the intent of UL standards to insure that products are "safe" when used as they are intended to be used and when abused to the point of failure that they shall not present a fire, electrical shock, or personal injury hazard.

A comprehensive understanding of the Standards that apply to a particular appliance is necessary to the design and production of the best appliance at the lowest price.

A design engineer would not design the electro-mechanical part of an appliance without first studying the applicable Standards. Very often, however, designers overspecify the material that is to be used for the enclosure. This happens because most designers do not adequately understand the requirements of the Standards that pertain to enclosure materials or the method of classifying polymeric materials under the UL Materials Recognition Program. Another feature of plastics that is often overlooked is their use as the support for the mechanical and electrical parts contained within the enclosure. Many UL Standards permit the use of plastic materials as the direct support for live electrical parts. An understanding of the Standards, especially UL 746-C (Polymeric Materials—

Use in Electrical Equipment Evaluations), allows the designer extra degrees of freedom by permitting design in which the functional parts of the appliance are directly supported by the enclosure.

Properties that are considered by UL in the evaluation of polymeric materials for suitability in electrical applications are divided into the two broad areas of short-term properties and long-term properties. These are covered in detail in UL Standards 746-A and 746-B, respectively.

Short-term properties include the following:

1. Mechanical, i.e., impact, flexural strength
2. Electircal, i.e., dielectric strength
3. Ease of ignition, i.e., arc ignition, hot wire ignition
4. Flammability, i.e., horizontal and vertical burn tests (UL Standard 94 covers this in detail)
5. Deflection under load, i.e., heat deflection temperature
6. Dimensional change under load, i.e., mold stress relief, creep
7. Chemical resistance

Long-term properties include the following:

1. Thermal endurance, i.e., relative thermal index (RTI)
2. Environmental exposure, i.e., UV, water

As an aid to the designer, the Underwriters Laboratories has established a "Materials Recognition Program," which pretests materials for the above properties. The most important properties are listed in the UL Recognized Component Directory and the remainder are available from the manufacturers upon request.

The designer will use the Recognized Component Directory, better known as the "Yellow Book," primarily to confirm the flammability rating, ignition rating, and RTI. Other pertinent properties can usually be more conveniently obtained from manufacturers' literature.

UL Standards are available from:

Underwriters Laboratories, Inc.
Publications Stock
333 Pfingsten Road
Northbrook, Illinois 60062
Telephone: 312/272-8800

VII. DESIGN TIP NUMBER SEVEN—WEATHERABILITY

Many end-use products in the course of normal service are exposed to the weather, either incidentally for short periods of time or continuously over long periods of time. The incidental type of exposure does not usually demand any particular consideration with regard to weathering effects on performance of the item. On the other hand, when products are likely to be subjected to extensive outdoor exposure, a prime consideration in selecting materials should be a concern for the weathering-resistance of the materials used.

All commonly used materials, such as wood, steel, and concrete are adversely affected by weather, and most thermoplastics show rather severe degradation in outside service. Consideration of Cycolac® ABS materials is recommended for any part or component where the initial properties would normally imply its suitability. It is the purpose of this design tip to describe the effect of weathering on Cycolac® brand ABS and to offer suggestions based on experience, to guide the user toward successful use of this material in outdoor application.

A. Effects of Weather on ABS

Prolonged exposure to the weather, and especially direct sunlight, will cause significant changes in both the appearance and the mechanical properties of ABS plastics.

B. Appearance or Aesthetic Changes

The material will lose gloss, shift in color tone toward yellow, and can surface-craze in areas of high strain. Very severe weather exposure in some climates can also degrade the surface to a chalking condition.

C. Changes in Mechanical Properties

The plastic will lose much of its impact resistance and ductility, particularly at low temperatures. Tensile and flexural strength values are maintained at normal temperatures, but these properties also drop appreciably at lower temperatures or higher strain rates. Modulus and hardness properties are not severely affected.

These changes in properties are due to the formation of a very thin, brittle layer on the exposed surface of the ABS. Any load sufficient to crack this veneer can, by "notch effect," cause the crack to propagate into the ductile core of the plastic.

D. Interpretation of Weathering Data

The primary purpose served by examining data on weathering of materials is the ability to predict reasonably the performance of a part made from those materials.

Most data on the aging behavior of plastics are acquired through accelerated tests and/or actual weather exposures. The primary purpose of this type of testing usually is to compare materials rather than to generate engineering design data. Comparisons between materials are made by measuring the retention of properties significant to the application as a function of exposure time (e.g., impact strength, gloss, tensile strength versus time).

In viewing the results of those accelerated methods, it is important to note carefully the conditions of exposure and test. Most studies on the subject are intended to collect as much data as possible using the smallest amount of sample possible, in the shortest possible period of time. This approach can lead to a severity of exposure and test data that bear little relationship to the actual use conditions that might be anticipated for an intended application.

In some cases, attempts to accelerate effects by artificial means can lead to erroneous and misleading results. In accelerated laboratory ultraviolet exposures, wavelengths of light are sometimes distributed differently than in normal sunlight, leading to effects not seen in outdoor weathering. Some methods can produce extremely high temperatures (over 200°F or 95°C) during exposures, again leading to nontypical effects. In actual outdoor tests, the standard procedure calls for specimen exposure on racks facing due south and at an angle of 45°. These are the conditions that offer a maximum exposure and intensity of direct sunlight. In a further attempt to accelerate outdoor effects, many studies are conducted in Arizona or Florida to obtain the most severe summer environments possible in the United States.

All of these accelerated tests are difficult to correlate with actual sunlight exposure. Even natural sunlight isn't "standard"; there are variations in clouds, smog, angle of the sun, rain, industrial environments, etc.

Relating exposure data from these types of tests to the actual performance requirements of the product must be done separately and specifically for each item by the manufacturer, in terms of design, contours, thickness, anticipated service conditions, expected outdoor exposure, and desired useful life of the product.

E. Outdoor Applications

Many outdoor applications in Cycolac® ABS have given satisfactory performance over a period of years because the designs were compatible with the performance requirements of the products involved. A realistic assessment of the product performance requirements is the first step to a successful application. Some items to be considered are:

1. Is it strictly an appearance part, or must it be expected to carry mechanical or thermal stresses?
2. Is the part likely to be subjected to fatigue stresses?
3. What will the storage and end-use environmental exposures be?
4. What is the required service life?

End-use products requiring exterior long-life retention of color, gloss, and abuse-resistance, should be designed with the surface protected with a pigmented protective film or coating.

F. Protective Paints

Specifically compounded paint systems based upon weather-resistant resins, having sufficient flexibility to avoid brittle-veneer effects, can also be effective in minimizing weather degradation.

G. Pigmentation

Pigmented Cycolac® ABS has somewhat better resistance to weathering that the natural unpigmented grades, both in appearance and in physical properties. When protective coatings cannot be used, a properly pigmented black grade should be chosen for best maintenance of physical properties.

Where colors must be used, the earthen tones are usually more satisfactory. Pastel blues or blue undertone colors which might shift toward green should be avoided. The best light-fast color pigments should be specified.

VIII. DESIGN TIP NUMBER EIGHT— RECOMMENDED TOLERANCES

Designers of plastic parts should be aware of the importance of specifying practical tolerances for their products. It is important that the designer carefully determine if the tolerances shown are realistic for the specified plastic and process. He should recognize that extreme accuracy of dimensions is expensive and, in some instances, impossible to hold in processing.

It is the responsibility of the designer to consider carefully these points, when specifying tolerances for the part design:

Are the specified tolerances, especially extremely close tolerances, necessary for the function of the part.
Indicate the conditions (i.e., temperature, time, environment, etc.), under which the specified dimensions must be held.
If extremely close tolerances are necessary, can the higher costs required for more precise tooling and processing be justified economically.
Only specific critical dimensions should have tight tolerances.

To assist the designer in this important area of tolerance standards, tables were prepared for the various plastics by the Custom Molders of the Society of Plastics Industry (see Tables 1–5A).

The tables should not be considered as hard and fast rules for all conditions, but should serve as a basis for establishing standards for molded products between designer-customer-molder.

Use of the Tables: The charts consist of two sets of values—commercial and fine. The commercial values or tolerances are those that can be held under normal processing conditions at the most economical level.

Fine values or tolerances are the narrowest possible limits of variation that can be held and only under controlled processing supervision using high-precision tooling.

IX. DESIGN TIP NUMBER NINE—MEASUREMENT OF THREE IMPORTANT DESIGN PROPERTIES

There are many physical properties with which designers, design engineers, and material specifiers are vitally concerned. In the pursuit of their everyday operations, they continually use physical properties to screen material for product applications. Of all the properties published in the plastics raw materials suppliers literature, information on three properties is most frequently requested. Most frequently requested properties are impact strength, heat deflection temperature, and flexural modulus. It is therefore concluded, that these properties are the most important tools in screening and specifying materials for applications.

A. Impact Strength

The Izod Impact Test, ASTM D256, is a widely used test for measuring the impact strength of plastics. A notched specimen bar mounted as a cantilevered beam is struck by a pendulum striker swung from a fixed height. As the pendulum swings down, breaking the sample, it continued through its arc, activating a pointer that gives a reading of the test. The test results are reported as foot-pounds per inch of notch.

Another test used to measure impact strength is the Falling Dart or Gardner Impact Test. A weighted "dart" with an impact nose of a specific radius is dropped onto the test specimen. The weight and height are also specified. The impact necessary to create a failure is then calculated via a series of tests and reported in foot pounds of energy.

The Falling Dart or Gardner Impact Test is considered to be more meaningful than the Izod Impact Test. Whereas these tests more closely approximate the dynamics of an end use impact, the Izod test more closely simulates a tearing test.

Although these tests do not as yet have an industry-wide testing specification, many companies have an internal standard method. Therefore, care must be taken when comparing the test results of various companies because of the likelihood of multiple variables. In view of this, we do not feel it advisable to show a comparison chart of these results.

B. Heat-Deflection Temperature

The Heat-Deflection Test, ASTM D648, indicates the point at which the tested plastic yields. This point is an arbitrary valve of deformation under the conditions of the test.

The test bar is supported underneath at both ends while immersed in a heated medium whose temperature is being raised continually. A rod exerts a load, in this test case at 264 psi, downward on the top center of the bar. A total deflection of 0.010 in. is the point at which the heat deflection temperature is reported.

C. Flexural Modulus

The ratio of flexural stress to flexural strain before permanent deformation has taken place is termed flexural modulus. The test for flexural modulus is ASTM D790.

The flexural strength of a material is the resistance to rupture under a bending stress applied at the center of a rectangular specimen supported at both ends. Flexural strength in psi is calculated from the load recorded at the breaking point and the cross-sectional area of the bar and the distance between supports. Effectively high flexural modulus values indicate a stiff material, whereas low values indicate a less rigid material.

SELECTED BIBLIOGRAPHY

Croft, Edward G., Quality Control in The Structural Foam Plant," Proceedings from the Ninth Structural Foam Conference, S.P.I., March 1981, pp. 67--72.

General Electric Co., Engineering Structural Foam Design Guide, General Electric Company, Plastics Operations, Pittsfield, Connecticut, 1983.

Hanson, Douglas I., Production, Economics, and Tooling, Proceedings from Eight Structural Foam Conference, S.P.I., 1980, pp. 80–85.

Hoover Universal, *Structural Foam Factbook*, Hoover Universal, Springfield, Massachusetts, 1982.

Mobay Chemical Corp., Snapfit Joints in Plastics, Mobay Chemical Corp., Pittsburgh, Pennsylvania, 1984.

Semerdjieu, Stefan, *Introduction to Structural Foam*, Society of Plastics Engineers, Inc., Brookfield Center, Connecticut, 1982.

The Society of the Plastics Industry, Inc., *Structural Foam*, The Society of the Plastics Industry, New York, 1982.

The Society of the Plastics Industry, Inc., *Structural Foam*, The Society of the Plastics Industry, Inc., New York, 1984.

Wendle, Bruce C., *Engineering Guide to Structural Foam*, Technomic Publishing Co., Inc., Westport, Connecticut, 1976.

Wendle, Bruce C., *Engineering Guide to Plastics Plant Layout and Machine Selection*, Technomic Publishing Co., Inc., Westport, Connecticut, 1978.

LIST OF APPENDIX TABLES

TABLE A.1 Standards and Practices of Plastics Molders: Engineering and Technical Standards ABS

STANDARDS AND PRACTICES OF PLASTICS MOLDERS

Engineering and Technical Standards
ABS

NOTE: The Commercial values shown below represent common production tolerances at the most economical level. The Fine values represent closer tolerances that can be held but at a greater cost.

Drawing Code	Dimensions (Inches)	Plus or Minus in Thousands of an Inch 1 2 3 4 5 6 7 8 9 10 11 12 13 14 15 16 17 18 19 20 21 22 23 24 25 26 27 28
A = Diameter (see Note #1)	0.000	
	0.500	
	1.000	
	2.000	Fine
B = Depth (see Note #3)	3.000	Commercial
	4.000	
C = Height (see Note #3)	5.000	
	6.000	

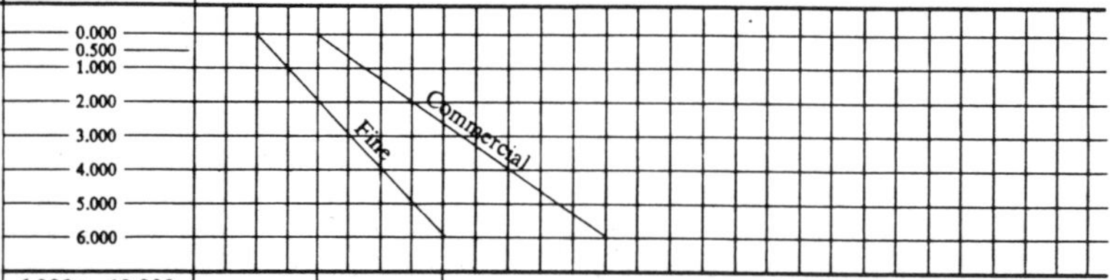

Drawing Code	Dimensions (Inches)	Comm. ±	Fine ±
	6.000 to 12.000 for each additional inch add (inches)	.003	.002
D=Bottom Wall (see Note #3)		.004	.002
E = Side Wall (see Note #4)		.003	.002
F = Hole Size Diameter (see Note #1)	0.000 to 0.125	.002	.001
	0.125 to 0.250	.002	.001
	0.250 to 0.500	.003	.002
	0.500 & Over	.004	.002
G = Hole Size Depth (see Note#5)	0.000 to 0.250	.003	.002
	0.250 to 0.500	.004	.002
	0.500 to 1.000	.005	.003
Draft Allowance per side (see Note #5)		2°	1°
Flatness (see Note #4)	0.000 to 3.000	.015	.010
	3.000 to 6.000	.030	.020
Thread Size (class)	Internal	1	2
	External	1	2
Concentricity (see Note #4)	(T.I.R.)	.009	.005
Fillets, Ribs, Corners (see Note #6)		.025	.015
Surface Finish	(see Note #7)		
Color Stability	(see Note #7)		

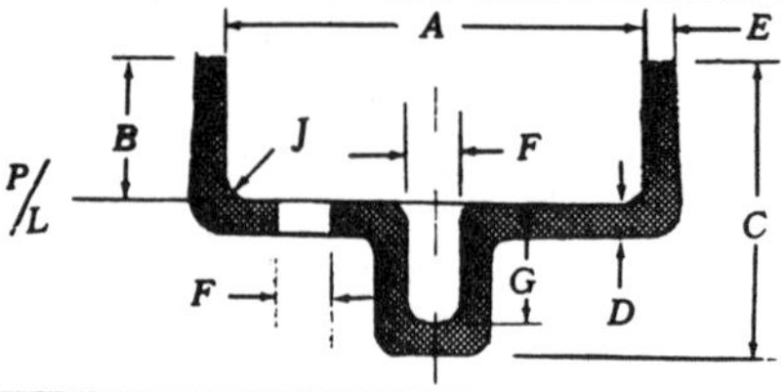

REFERENCE NOTES

1 – These tolerances do not include allowance for aging characteristics of material.

2 – Tolerances based on ⅛" wall section.

3 – Parting line must be taken into consideration.

4 – Part design should maintain a wall thickness as nearly constant as possible. Complete uniformity in this dimension is impossible to achieve.

5 – Care must be taken that the ratio of the depth of a cored hole to its diameter does not reach a point that will result in excessive pin damage.

6 – These values should be increased whenever compatible with desired design and good molding technique.

7 – Customer-Molder understanding necessary prior to tooling.

Source: The Society of The Plastics Industry, Inc.

TABLE A.2 Standards and Practices of Plastics Molders: Engineering and Technical Standards Polycarbonate

STANDARDS AND PRACTICES OF PLASTICS MOLDERS

Engineering and Technical Standards
POLYCARBONATE

NOTE: The Commercial values shown below represent common production tolerances at the most economical level. The Fine values represent closer tolerances that can be held but at a greater cost.

Drawing Code	Dimensions (Inches)	Plus or Minus in Thousands of an Inch 1 2 3 4 5 6 7 8 9 10 11 12 13 14 15 16 17 18 19 20 21 22 23 24 25 26 27 28	
A = Diameter (see Note #1) B = Depth (see Note #3) C = Height (see Note #3)	0.000 0.500 1.000 2.000 3.000 4.000 5.000 6.000	Fine	Commercial
	6.000 to 12.000 for each additional inch add (inches)	Comm. ±	Fine ±
		.003	.0015
D=Bottom Wall (see Note #3)		.003	.002
E = Side Wall (see Note #4)		.003	.002
F = Hole Size Diameter (see Note #1)	0.000 to 0.125	.002	.001
	0.125 to 0.250	.002	.0015
	0.250 to 0.500	.003	.002
	0.500 & Over	.003	.002
G = Hole Size Depth (see Note#5)	0.000 to 0.250	.002	.002
	0.250 to 0.500	.003	.002
	0.500 to 1.000	.004	.003
Draft Allowance per side (see Note #5)		1°	½°
Flatness (see Note #4)	0.000 to 3.000	.005	.003
	3.000 to 6.000	.007	.004
Thread Size (class)	Internal	1B	2B
	External	1A	2A
Concentricity (see Note #4)	(T.I.R.)	.005	.003
Fillets, Ribs, Corners (see Note #6)		.015	.015
Surface Finish	(see Note #7)		
Color Stability	(see Note #7)		

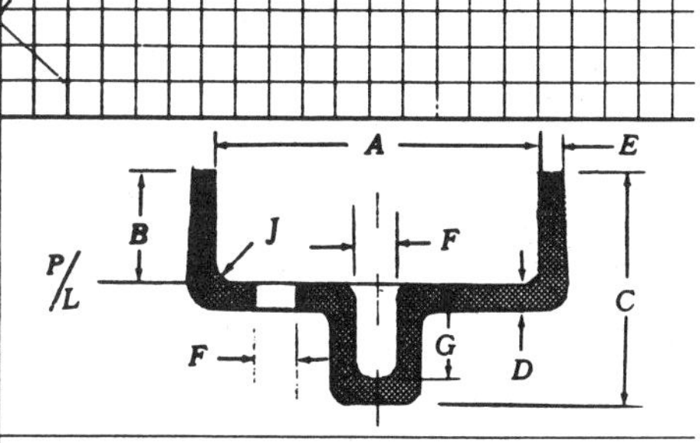

REFERENCE NOTES

1 – These tolerances do not include allowance for aging characteristics of material.

2 – Tolerances based on ⅛" wall section.

3 – Parting line must be taken into consideration.

4 – Part design should maintain a wall thickness as nearly constant as possible. Complete uniformity in this dimension is impossible to achieve.

5 – Care must be taken that the ratio of the depth of a cored hole to its diameter does not reach a point that will result in excessive pin damage.

6 – These values should be increased whenever compatible with desired design and good molding technique.

7 – Customer-Molder understanding necessary prior to tooling.

Source: The Society of The Plastics Industry, Inc.

TABLE A.3 Standards and Practices of Plastics Molders: Engineering and Technical Standards High Density Polyethylene

STANDARDS AND PRACTICES OF PLASTICS MOLDERS

Engineering and Technical Standards
HIGH DENSITY POLYETHYLENE

NOTE: The Commercial values shown below represent common production tolerances at the most economical level. The Fine values represent closer tolerances that can be held but at a greater cost.

Drawing Code	Dimensions (Inches)	Plus or Minus in Thousands of an Inch 1 2 3 4 5 6 7 8 9 10 11 12 13 14 15 16 17 18 19 20 21 22 23 24 25 26 27 28
A = Diameter (see Note #1)	0.000 0.500 1.000 2.000	Commercial Fine
B = Depth (see Note #3)	3.000 4.000	
C = Height (see Note #3)	5.000 6.000	

Drawing Code	Dimensions (Inches)	Comm. ±	Fine ±
	6.000 to 12.000 for each additional inch add (inches)	.006	.003
D=Bottom Wall (see Note #3)		.006	.004
E = Side Wall (see Note #4)		.006	.004
F = Hole Size Diameter (see Note #1)	0.000 to 0.125	.003	.002
	0.125 to 0.250	.005	.003
	0.250 to 0.500	.006	.004
	0.500 & Over	.008	.005
G = Hole Size Depth (see Note#5)	0.000 to 0.250	.005	.003
	0.250 to 0.500	.007	.004
	0.500 to 1.000	.009	.006
Draft Allowance per side (see Note #5)		2°	¾°
Flatness (see Note #4)	0.000 to 3.000	.023	.015
	3.000 to 6.000	.037	.022
Thread Size (class)	Internal	1	2
	External	1	2
Concentricity (see Note #4)	(T.I.R.)	.027	.010
Fillets, Ribs, Corners (see Note #6)		.025	.010
Surface Finish	(see Note #7)		
Color Stability	(see Note #7)		

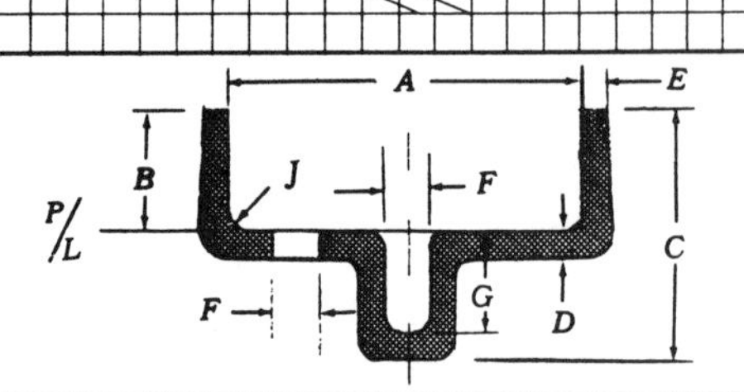

REFERENCE NOTES

1 – These tolerances do not include allowance for aging characteristics of material.

2 – Tolerances based on ⅛" wall section.

3 – Parting line must be taken into consideration.

4 – Part design should maintain a wall thickness as nearly constant as possible. Complete uniformity in this dimension is impossible to achieve.

5 – Care must be taken that the ratio of the depth of a cored hole to its diameter does not reach a point that will result in excessive pin damage.

6 – These values should be increased whenever compatible with desired design and good molding technique.

7 – Customer-Molder understanding necessary prior to tooling.

Source: The Society of The Plastics Industry, Inc.

TABLE A.4 Standards and Practices of Plastics Molders: Engineering and Technical Standards Polypropylene

STANDARDS AND PRACTICES OF PLASTICS MOLDERS

Engineering and Technical Standards
POLYPROPYLENE

NOTE: The Commercial values shown below represent common production tolerances at the most economical level. The Fine values represent closer tolerances that can be held but at a greater cost.

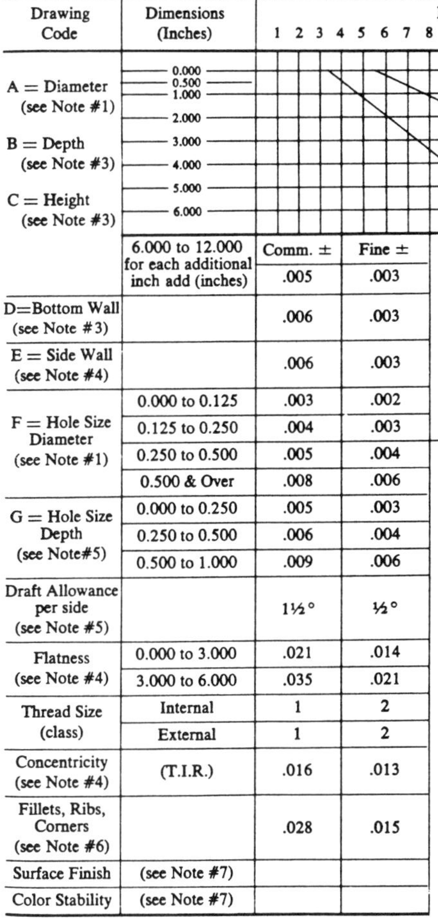

Drawing Code	Dimensions (Inches)		
A = Diameter (see Note #1)	0.000		
	0.500		
	1.000		
	2.000		
B = Depth (see Note #3)	3.000		
	4.000		
C = Height (see Note #3)	5.000		
	6.000		
	6.000 to 12.000 for each additional inch add (inches)	**Comm. ±**	**Fine ±**
		.005	.003
D=Bottom Wall (see Note #3)		.006	.003
E = Side Wall (see Note #4)		.006	.003
F = Hole Size Diameter (see Note #1)	0.000 to 0.125	.003	.002
	0.125 to 0.250	.004	.003
	0.250 to 0.500	.005	.004
	0.500 & Over	.008	.006
G = Hole Size Depth (see Note#5)	0.000 to 0.250	.005	.003
	0.250 to 0.500	.006	.004
	0.500 to 1.000	.009	.006
Draft Allowance per side (see Note #5)		1½°	½°
Flatness (see Note #4)	0.000 to 3.000	.021	.014
	3.000 to 6.000	.035	.021
Thread Size (class)	Internal	1	2
	External	1	2
Concentricity (see Note #4)	(T.I.R.)	.016	.013
Fillets, Ribs, Corners (see Note #6)		.028	.015
Surface Finish	(see Note #7)		
Color Stability	(see Note #7)		

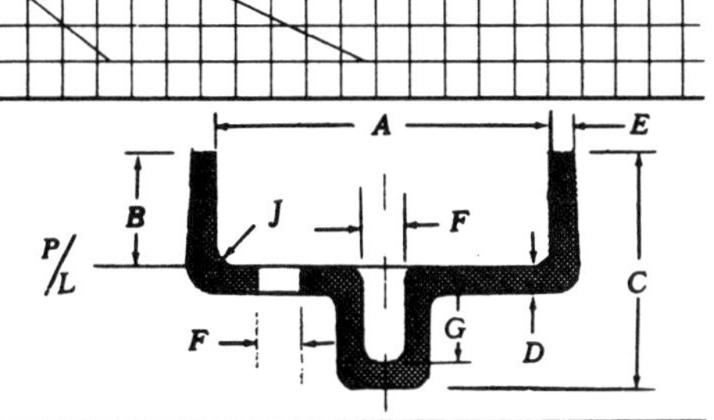

REFERENCE NOTES

1 – These tolerances do not include allowance for aging characteristics of material.

2 – Tolerances based on ⅛" wall section.

3 – Parting line must be taken into consideration.

4 – Part design should maintain a wall thickness as nearly constant as possible. Complete uniformity in this dimension is impossible to achieve.

5 – Care must be taken that the ratio of the depth of a cored hole to its diameter does not reach a point that will result in excessive pin damage.

6 – These values should be increased whenever compatible with desired design and good molding technique.

7 – Customer-Molder understanding necessary prior to tooling.

Source: The Society of The Plastics Industry, Inc.

TABLE A.5 Standards and Practices of Plastics Molders: Engineering and Technical Standards Polystyrene

STANDARDS AND PRACTICES OF PLASTICS MOLDERS

Engineering and Technical Standards
POLYSTYRENE

NOTE: The Commercial values shown below represent common production tolerances at the most economical level. The Fine values represent closer tolerances that can be held but at a greater cost.

Drawing Code	Dimensions (Inches)	Plus or Minus in Thousands of an Inch 1 2 3 4 5 6 7 8 9 10 11 12 13 14 15 16 17 18 19 20 21 22 23 24 25 26 27 28
A = Diameter (see Note #1)	0.000	
	0.500	
	1.000	
	2.000	Fine
B = Depth (see Note #3)	3.000	Commercial
	4.000	
	5.000	
C = Height (see Note #3)	6.000	

Drawing Code	Dimensions (Inches)	Comm. ±	Fine ±
	6.000 to 12.000 for each additional inch add (inches)	.004	.002
D=Bottom Wall (see Note #3)		.0055	.003
E = Side Wall (see Note #4)		.007	.0035
F = Hole Size Diameter (see Note #1)	0.000 to 0.125	.002	.001
	0.125 to 0.250	.002	.001
	0.250 to 0.500	.002	.0015
	0.500 & Over	.0035	.002
G = Hole Size Depth (see Note#5)	0.000 to 0.250	.0035	.002
	0.250 to 0.500	.004	.002
	0.500 to 1.000	.005	.003
Draft Allowance per side (see Note #5)		1½°	½°
Flatness (see Note #4)	0.000 to 3.000	.007	.004
	3.000 to 6.000	.013	.005
Thread Size (class)	Internal	1	2
	External	1	2
Concentricity (see Note #4)	(T.I.R.)	.010	.008
Fillets, Ribs, Corners (see Note #6)		.015	.010
Surface Finish	(see Note #7)		
Color Stability	(see Note #7)		

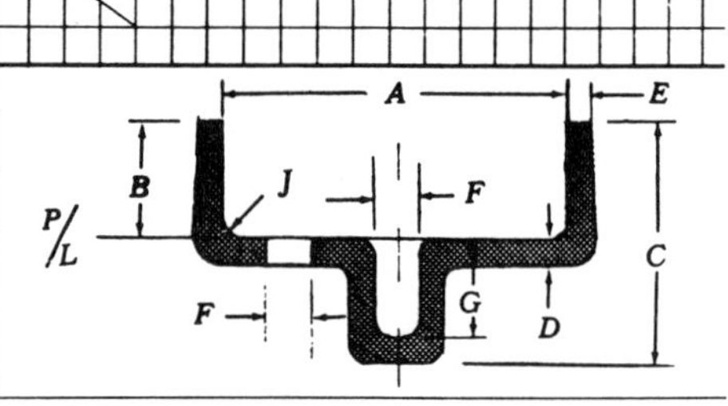

REFERENCE NOTES

1 – These tolerances do not include allowance for aging characteristics of material.

2 – Tolerances based on ⅛″ wall section.

3 – Parting line must be taken into consideration.

4 – Part design should maintain a wall thickness as nearly constant as possible. Complete uniformity in this dimension is impossible to achieve.

5 – Care must be taken that the ratio of the depth of a cored hole to its diameter does not reach a point that will result in excessive pin damage.

6 – These values should be increased whenever compatible with desired design and good molding technique.

7 – Customer-Molder understanding necessary prior to tooling.

Source: The Society of The Plastics Industry, Inc.

Index